Know Thy Adversary but first and foremost Know Thyself

Battle Tested Strategies From The Front Lines For CISOs.

By

DEVON BRYAN

CISSP, CIPP E/U

ii

Contents

Preface

Although written more than 2500 years ago, many of the critical teachings and techniques outlined in Sun Tzu's "Art of War" have proven directly relevant to the many flavors of modern warfare, with cybersecurity being no exception.

Among my favorite Sun Tzu's battlefield lessons is the importance of knowledge and insights into one's capabilities, motivations, and the intent of one's adversaries. This is paraphrased as, *"If you know the enemy and know yourself, you need not fear the result of a hundred battles. If you know yourself but not the enemy, for every victory gained, you will also suffer a defeat. If you know neither the enemy nor yourself, you will succumb in every battle."* - Sun Tzu.

Show me a practicing Chief Information Security Officer (CISO) whose operational strategies do not embody this principle. I will show you a CISO whose career will undoubtedly be cut short by a significant breach or three.

Foreword

I t was late September 2001, and there I was, standing in the middle of the Internal Revenue Service's (IRS) newly created Computer Security Incident Response Center (CSIRC). The system status indicators on the overhead monitors progressively changed colors from Green (Normal) to Red (Critical) as the NIMDA worm propagated from system to system, network to network. This was infecting the IRS' networks and systems as it progressed.

Released precisely seven days after the tragic attacks of September 11, 2001, NIMDA (admin spelled backward) was a hybrid worm that spread via infected email attachments and across websites running vulnerable versions of Microsoft's Internet Information Services (IIS) web server software. The malware exploited a folder traversal vulnerability, which was only patched by Microsoft a month after the initial outbreak on September 18, 2001.

The infuriating program infected various sites worldwide, causing significant operational problems mainly due to its aggressive spreading techniques. The worm exploited weak passwords to speed across different machines on local networks and propagated using backdoors left open after the Code Red II worm outbreak.

The Code Red II worm was a significant cyber threat that emerged shortly after the original Code Red worm in 2001. Code Red II was a variant of the original Code Red worm, designed to exploit vulnerabilities in Microsoft's IIS web servers. It specifically targeted a buffer overflow vulnerability in the IIS indexing service. Unlike its predecessor, Code Red II was primarily a backdoor Trojan rather than a simple web defacement worm. Once it compromised a server, it created a backdoor, which allowed attackers to access and control the infected machine remotely. While the original Code Red worm focused on spreading rapidly and defacing websites, Code Red II's aim was more about establishing control over compromised systems, potentially launching further attacks, or building a network of "zombie" machines. Code Red II randomly selects Internet Protocol (IP) addresses to scan and infect other vulnerable IIS servers. Once infected, a machine would become part of a more extensive network that could be used for Distributed Denial-of-Service (DDoS) attacks or other malicious purposes. This worm affected thousands of servers, causing significant disruptions. Using backdoors in compromised machines also raised concerns about potential data breaches and unauthorized access to sensitive information.

And there I was, relatively fresh out of an 11-year career in the United States Air Force (USAF) and a 3-year stint as a DC Beltway Contractor. Afterwards, I'd landed the gig as the IRS' Associate Director of Incident Response, which made me responsible for preventing, detecting, and responding to cyber-attacks against the agency's systems.

If Helmut von Moltke (renowned military tactician) was correct that *"No battle plan ever survives first contact with the enemy,"* what determines a leader's natural and intuitive reactions in crises?

In my case, what was it about my education, training, and life experiences to date that prepared me for this moment of operational crisis? What leadership lessons from my five-year stint as a USAF Commissioned Military Officer, or my six years as a Supply Technician prepared me for this moment when split-second decision-making could lead to massive consequences?

Arguably, that task boils down to initiating the appropriate and timely containment, eradication, and recovery techniques. This will prevent further damage to uninfected IRS systems; ensure operational resilience; and quickly restore infected systems to full operational capability. Furthermore, what impact, if any, would this singular incident have on shaping my future career as a CISO?

This book, born out of years of experience on the frontlines of cybersecurity, explores those lessons—lessons forged in the heat of real-world crises, refined through decades of practice and grounded in Sun Tzu's timeless strategies. As you turn these pages, you will journey through the critical moments that shaped my approach to cybersecurity, from the cyber trenches of the USAF communications systems and networks to the boardrooms of Fortune 500 companies.

In these chapters, you'll find the strategies and frameworks that have guided me and the underlying principles that every CISO and cyber defender should internalize in some form or another.

We live in an era where the battlefield of global business has become increasingly digital, and cyber adversaries are as cunning as they are relentless. Yet, as Sun Tzu reminds us, the path to victory lies in thorough preparation, deep understanding, and unwavering vigilance.

Whether you're an experienced CISO or new to the field, I hope this book will serve as a guide and a source of inspiration. It will help you navigate the complex and often tricky cybersecurity landscape with confidence, clarity, and a relentless commitment to protecting what matters most.

Context is Everything

As seasoned cybersecurity practitioners know, 'context is everything' regarding our tradecraft. As a backdrop to the rest of this story, it's essential that I provide you with some information about the significant events and turning points that have collectively helped shape my career trajectory to the pinnacle role in the cyber security profession.

My Chief Information Security Officer (CISO) journey began with my boyhood interest in medicine. This landed me in the Natural Sciences program at the University of the West Indies in Kingston, Jamaica, where I excelled in Mathematics. I later joked that my early interest in curing human viruses and infections had translated to my eventual career as a cyber defender protecting corporations from computer viruses.

Soon after migrating to the United States of America from the tiny island of Jamaica in July of 1987, I enlisted into the U.S. Air Force in January of 1988 to serve our great democracy while earning benefits towards college tuition. Among the handful of USAF active-duty jobs available to non-US citizens, the position of a Supply Technician landed me on a dreaded and scorned graveyard shift. I worked from midnight to 8AM as the 'newbie' in the Supply Squadron at Ellsworth Air Force Base in Rapid City South Dakota.

After spending the first two years of my enlistment getting accustomed to my military life and third shift duties, I enrolled as a full-time student in the Applied Sciences Department of the only local engineering school. This was South Dakota School of Mines and Technology (SDSM&T) and I majored in Applied Mathematics and Computer Science.

Between a rigorous academic curriculum at SDSM&T; my commitment at home as a husband and a new dad; and my graveyard shift duties for the USAF, which coincidentally also exposed me to the innards of the Sperry-Univac 1100/60 that was running the USAF's inventory management system at the time; my days and nights were thoroughly filled.

Upon graduation from SDSM&T in December of 1992 with a Bachelor of Science in Applied Mathematics and a minor in Computer Science, I applied to the Air Force's Commissioned Officer Training Program. I was blessed and lucky enough to be accepted. What was true back then, and is still accurate today, is that the U.S. Armed Forces, the Air Force being no exception, desperately need computer scientists. The military was maximizing its usage of information technology and realized its broader implications in modern warfare, kinetic and non-kinetic.

Notably, none of the courses I had to take for my degree in school were related to computer security, which is understandable based on the time-period involved. From a practitioner's perspective, however, considering that only four years prior, the first major network-borne malware, the Morris Worm, ran rampant across the Internet, undergraduate collegiate computer sciences curricula should really have begun catching on to the longer-term ramifications of network-borne badness but hadn't yet.

For additional context, the Morris Worm, released in 1988, is often considered the first widely recognized computer worm and one of the earliest significant events in cybersecurity history. Created by Robert Tappan Morris, a graduate student at Cornell University, the worm was intended as an experiment to measure the size of the Internet. However, a coding flaw caused it to spread uncontrollably. The worm exploited known vulnerabilities in Unix systems, including weak passwords and deficiencies in several programs (like Sendmail and Finger). It spreads by infecting systems and replicating itself across networks. Morris designed the worm to avoid repeatedly infecting the same machine, but a bug in the code led to repeated infections. As a result, infected systems became overwhelmed by excessive resource usage, effectively shutting them down. The worm infected around 6,000 machines, approximately 10% of the Internet at the time, causing widespread network slowdowns and system outages. It highlighted vulnerabilities in widely used systems, leading to significant administrative cleanup costs. This incident marked one of the first cases of prosecuting a computer crime. Robert Morris was charged under the Computer Fraud and Abuse Act and sentenced to probation, community service, and a fine.

Upon graduating from basic commissioned officer training, I was sent to Basic Communications Officer Training School. I received USAF-tailored training in network management, advanced computer programming, and in-depth training on systems supporting command and control operations there. Following this four-month-long course, I reported for my first duty station at Falcon Air Force Base Colorado Springs as a communications officer. I was assigned to the 4th Space Operations Squadron, which was charged with engineering responsibilities for the communications payload onboard the Milstar Satellite System.

Milstar (Military Strategic and Tactical Relay) is operated by the USAF. It was designed to provide secure and jam-resistant worldwide communications for the U.S. Military. Each satellite served as a direct traffic between terminals on the Earth. Designed to deliver communications that are hard to detect and intercept and survivable in the event where the satellites process the signals transmitted to them. They can link with other Milstar satellites through crosslinks, reducing the ground-controlled switching requirement.

My USAF Communications Officer training and the depth and breadth of exposure during my first duty assignment were instrumental in shaping my career as a cybersecurity practitioner. Ranking high among the many career-changing influences of my three-year tenure at Falcon AFB was that my first commanding officer was the then Major Suzanne Vautrinot. Major Suzanne Vautrinot was the Major General and Commander, 24th Air Force, Air Forces Cyber and Air Force Network Operations and Director of Plans and Policy, U.S. Cyber Command and Deputy Commander, Network Warfare, U.S. Strategic Command.

However, what left an indelible mark on me and my career was the relationship between secure communications and systems to mission success. This is especially important when lives are on the line and national security matters are critical. These lessons were carried over to my next assignment as a Network Security Engineer for the Air Combat Command's Computer Systems Squadron and Communications Group at Langley AFB in Hampton Roads, Virginia where I was among the initial cadre of young lieutenants charged with deploying early generation firewalls (Sidewinder firewalls from McAfee/Intel) and using the intrusion detection systems Automated Security Measurement System (ASIM) to monitor network traffic of the USAF Air Combat Command.

Little did I know at the time that my career as a cyber defender was overlapping with the history of intrusion detection systems. This began in earnest in the 1990s and saw that the Air Force's Cryptologic Support Center development of the Automated Security Measurement System (ASIM) would monitor network traffic on the U.S. Air Force. ASIM made considerable progress in overcoming scalability and portability issues that previously problem-plagued other Network Intrusion Detection Systems (NIDS) products.

Additionally, ASIM was the first solution to incorporate both a hardware and software solution to network intrusion detection. ASIM was being managed by the USAF Information Warfare Squadron at locations worldwide. As often happens, the development group on the ASIM project, which included several former USAF active-duty airmen, formed a commercial company in 1994, the WheelGroup. Their product, NetRanger, was the first commercially viable network intrusion detection device. The intrusion detection market began to gain popularity and generate revenues around 1997. That year, the security market leader, Internet Security Systems (ISS), developed a network intrusion detection system called RealSecure. A year later, Cisco recognized the importance of network intrusion detection and purchased the WheelGroup, attaining a security solution they could provide their customers.

In my second assignment, I was fortunate to have as my Squadron Commander, then Lieutenant Colonel (Lt Col) Ronnie Hawkins, and Base Communications Commander, then Colonel (Col) Dale Meyerrose. These great leaders both went on to have very accomplished USAF careers with Lt Col. Hawkins attaining the rank of Lieutenant General and Director, Defense Information Systems Agency and Commander, Joint Force Headquarters – Department of Defense Information Networks (DODIN) in Fort Meade, Maryland. Col Meyerrose also went on to become a Lieutenant General and the first President-appointed, Senate-confirmed Associate Director of National Intelligence/Intelligence Community Chief Information Officer (CIO) and Information Sharing Executive for the Office of the Director of National Intelligence (ODNI).

Major General Meyerrose was among the early military strategists who recognized the importance of dominance of the cyber domain as the next domain of warfare. His early briefings on "The Network as a Weapons System" and the creation of Network Operations and Security Centers (NOSCs) were forerunners to what has since become the U.S. Cyber Command.

In addition to the many lessons from the two great leaders I was privileged enough to support at Langley AFB, my second assignment was further enriched by two other events that significantly influenced my career trajectory. First was my nine-month deployment to Saudi Arabia, supporting Operation Southern Watch. Second and perhaps of most pertinence was my peripheral involvement in what has been broadly called one of the most aggressive and direct cyber-attacks against the U.S. Military.

Operation Southern Watch began on August 27, 1992, to ensure Iraqi compliance which demanded that Iraq end its repression and ensure that the human and political rights of all Iraqi citizens are respected. The end of Iraq bombing and strafing attacks against the Shi'ite Muslims in Southern Iraq during the remainder of 1991 and into 1992 indicated those that chose not to comply with the resolution. Military forces from the USA, Canada and the United Kingdom participated in Operation Southern Watch under the command of the Joint Task Force Southwest Asia Commander (JTF SWA), who reported directly to the Commander (USCENTCOM).

My team of communications personnel landed in Riyadh, Saudi Arabia, in the Spring of 1994 with orders to deliver voice, video, and data communications to warfighters operating out of the growing number of tent cities sprinkled all over Saudi Arabia and Kuwait. During my nine months of hauling military caravans loaded with communications equipment to various forward operating bases, acquiring satellite uplink/downlink engineering and configuring secure/insecure voice, video, and data communications, there were many more essential career lessons on the importance of communications to successful military campaigns encapsulated into the slogan often proudly displayed by tactical communications units, "No Comms, No Bombs" to underscore the importance of their contributions to the mission.

My second career-impactful lesson from my Langley AFB rotation came from the January 1997 Distributed Denial of Service (DDoS) attacks targeting the Base's email servers, widely reported in the media and academia as the "Langley Cyber Attack." Miscreants abusing flaws in the Air Force Base's email relay servers to send spoofed emails to unsuspecting targets became highly upset with the local team's discovery of their activities and the early countermeasures deployed to deny them access actively. In direct retaliation to the early countermeasures implemented, the miscreants flooded the Base's email infrastructures with unprecedented traffic volumes.

The Black Hole Strategy adopted by the Incident Response Team for future cyber-attacks was a crude rules-based email bomb filter cited by scholars as the first documented "crude spam filter," predating today's sophisticated ones. The Langley AFB Cyber Attack was a seminal event not just because of its strategic cybersecurity implications for the military and industry alike, but also because it was perhaps the most prominent and earliest "Know Thyself" lesson of my career as a cyber practitioner. The fact that this component of the external cyber-attack surface of one of the USAF Air Combat Command's most important bases was running a misconfigured and easily exploitable version of Sendmail became a haunting motivation for me to always to ensure I'm implementing programmatic controls to instrument the external perimeter of the organizations I'm charged with protecting from an external attack surface management perspective.

After my promotion to the rank of Captain, my duties and responsibilities began to change with USAF career advancement norms. For Commissioned Officers to progress in rank, their duties tend to become increasingly more administrative and far less technical. Therefore, I faced a significant career-changing decision. As much as I loved serving our great democracy as an armed forces member, my passion for being close to the technologies enabling cyber defense drove me to separate from the USAF.

In January 1999, I was honorably discharged from the USAF after serving for eleven years. I was awarded numerous citations and medals for participating in various military campaigns and amassed several leading technology industry certifications including Cisco, Microsoft, and Computer Associates. All while completing a Master of Science Degree in Computer Science from Colorado Technological University in Colorado Springs, Colorado.

Reflecting on my journey, it becomes clear that my career trajectory—from a young boy with a passion for medicine to a seasoned cybersecurity professional—was shaped by a series of pivotal experiences and influential mentors. The path was neither straight nor easy, but each challenge, from the rigors of military life to the complexities of early cyber defense, prepared me for the multifaceted role of a CISO.

This journey has been one of constant learning, adaptation, and growth. And while the landscape of cybersecurity continues to evolve, one thing remains certain: the need for vigilant, informed, and strategic defenders is more critical than ever. As we move forward into new challenges and frontiers, may the lessons from the past guide us, and the wisdom we gain today shape tomorrow's battles.

Chapter 1

KNOW THYSELF...

Foundational Building Blocks

B efore heading to the IRS to lead the creation of their first Computer Security Incident Response Center (CSIRC) as a consultant working for SRA International (now CSRA), I spent the first few years of my civilian life deploying firewalls and intrusion detection systems for several federal government agencies and Washington DC law firms as a consultant working for Wang Government Systems and Mainstay Enterprises.

As a young consultant to the IRS, Sun Tzu's "Know Thyself" had early implications for me beyond the typical daily 'blocking and tackling' of cyber threats, aka cyber trench warfare that we engage in as cyber defenders. At the IRS, as is true for most large organizations, the power of information technology was being harnessed to transform how the agency delivers its services to U.S. taxpayers.

It has indeed been said that one of the great paradoxes of our information age is that the same Internet technologies that bring so much positive transformation bring with them so much peril from cyber miscreants seeking to subvert, disrupt, and exploit internet-connected systems. As is true nowadays, with so many other aspects of our personal and professional lives, the market-growth strategies of modern businesses, including the IRS, have become increasingly IT-centric. Hence, synonymous with the constant processing of staggering volumes of susceptible information in real-time across intricately connected high-speed data networks, has resulted in the consequent expansion of cyber-attack surfaces creating ample targets for bad actors.

Thanks to my early exposure, I came to quickly realize what is now of vital importance for cyber defenders' irrespective of industry, that we are not in the business of cybersecurity for the sake of cybersecurity but because we're employed to support and enable our organizations' significant lines of businesses in their go-to-market strategies. This requires a deliberate focus to ensure cybersecurity strategies align with and enhance business strategies. It also requires that cyber defenders drive the right balance between threat-driven, risk-based trade-offs to ensure that cybersecurity programs; policies; and strategies enable businesses to go further and faster. Thankfully, these lessons were emblazoned in me early enough as core competencies that I must also 'Know Thy Businesses' as a core component of "Knowing Thyself."

Among the many "Know Thyself" questions swirling in a cyber defender's mind at the onset of cyber hostilities are the infection method and propagation techniques of the malware variant they're battling.

These other questions include:
- Do we know what's directly connected to our systems and networks?
- Do we know what software is running (or trying to run) on our systems and networks?
- Do we know what our critical business functions and processes are?
- What applications, systems, and networks support those critical business functions?

- Do we maintain a current inventory of vulnerable systems by type, location, and criticality level to the organization's mission?
- Have we sufficiently instrumented the surface of our external cyberattack threat?
- Do we maintain a current inventory of containment mechanisms that can be quickly activated in response to a cyberattack?
- Do we know which of our employees and contractors have administrative access to what systems and for what reasons?

Ideally, cyber defenders would have gathered much of this information about their organization well before a disruptive cyber event threatened them.

We should know what's connected to our systems' networks and actively manage (inventory, track, and correct) all hardware devices and software assets on the network so that only authorized devices are given access and unauthorized and unmanaged devices are found and prevented from gaining access.

Know Thy APIs

The Importance of CISOs Knowing Their API Landscape Know Thy APIs

And while we're in the software stack, CISOs should also have a keen sense of their API or Application Programming Interface landscape. APIs are typically used to access web-based services, operating system features, or libraries. They enable seamless communication between different software systems by specifying the methods and data format applications that should be used when making requests and receiving responses. They are therefore instrumental in fostering interoperability, allowing diverse applications to work together efficiently and facilitating the development of robust, modular software.

Managing an API inventory is therefore crucial for the same reasons we've discussed maintaining hardware assets, software assets, and SBOM inventories, which include:

- **Visibility and Transparency:** Maintaining an API inventory provides clear visibility into the APIs used within an organization. It allows developers, architects, and business stakeholders to understand what APIs are available, their functionalities, and how they are utilized across different systems and applications.

- **Security and Compliance:** Understanding the APIs in use is vital for managing security risks. API vulnerabilities can be potential entry points for cyber-attacks. Organizations can identify security risks associated with specific APIs and implement appropriate security measures by maintaining an API inventory. Additionally, in regulated industries, having a clear API inventory is essential for compliance purposes, ensuring that data is handled securely and per relevant regulations.

- **Vendor Management**: Many organizations use third-party APIs from external vendors or partners. Managing an API inventory helps in vendor management; enabling organizations to track service level agreements (SLAs); understand dependencies on external APIs; and assess the impact of vendor-related changes on internal systems.

In summary, access to a robust API inventory is crucial for cyber defenders to better protect all critical components of their organization's software ecosystems as a comprehensive-attack, surface management strategy subcomponent.

Know Thy SaaS Footprint

Know Thy SaaS Footprint - The New Perimeter in the Cloud Era As organizations increasingly move toward cloud-based solutions, Software as a Service (SaaS) has become integral to business operations. SaaS applications enable scalability, flexibility, and cost-effectiveness, transforming how companies manage workflows, collaborate, and store data. However, as the SaaS footprint grows, so does the organization's attack surface. For CISOs, understanding and managing the SaaS footprint is essential for maintaining security, compliance, and control over sensitive data. Organizations can avoid shadow

The Expanding SaaS Footprint: A Modern-Day Business Necessity

Data breaches, and regulatory non-compliance with clear visibility into SaaS applications.

This chapter explores why knowing the SaaS footprint is crucial for CISOs, the security challenges SaaS introduces, and practical strategies for monitoring, governing, and securing SaaS usage across the enterprise.

SaaS applications offer organizations the agility and speed to stay competitive, reducing infrastructure management and enabling global collaboration. SaaS usage spans nearly all business functions, from H.R. and finance to sales and customer support. However, this expansion has brought new complexities, and without adequate oversight, organizations risk the following:

1. **Data Security Issues:** SaaS platforms store sensitive and proprietary information, from customer data to intellectual property.
2. **Compliance Challenges:** Regulatory requirements such as GDPR, HIPAA, and CCPA require stringent data privacy practices that apply to all third-party platforms.

3. **Access Control Concerns:** Inadequate Identity and Access Management (IAM) for SaaS applications can lead to unauthorized access and privilege creep.

4. **Shadow IT:** Employees often adopt SaaS tools independently, creating unknown and unmanaged access points in the organization's IT environment.

In light of these challenges, CISOs must understand which SaaS applications are in use and how they are used, and by whom, and with what security controls are in place.

Why CISOs Must Prioritize SaaS Visibility

Visibility into the SaaS footprint allows CISOs to proactively manage risks, ensure compliance, and support secure innovation. Key reasons CISOs must prioritize SaaS visibility include:

1. **Preventing Data Breaches and Loss:** SaaS applications often handle sensitive information, such as Personally Identifiable Information (PII) and proprietary data. Data loss or unauthorized access can occur without proper monitoring, resulting in significant reputational and financial damage.

2. **Controlling Shadow IT:** Employees and departments may deploy SaaS tools without IT approval. Shadow IT creates security blind spots where sensitive data may reside or transfer without the necessary security measures.

3. **Ensuring Compliance:** SaaS platforms must meet data protection and privacy regulatory requirements. CISOs must ensure that each SaaS provider complies with relevant regulations and offers adequate security measures.

4. **Managing Access and Privileges:** Excessive privileges in SaaS applications can lead to unauthorized access, so identity and access management is crucial for every SaaS tool.

5. **Mitigating Vendor and Supply Chain Risk:** Each SaaS provider represents a potential risk in the supply chain. Ensuring that SaaS vendors implement strong security mea-

sures and follow best practices reduces the risk of third-party breaches.

Without this visibility, organizations cannot enforce security policies effectively, risking compromised data and regulatory non-compliance.

Common Security Risks Associated with SaaS

SaaS applications can expose organizations to a range of security risks, including:

1. **Data Leakage:** Sensitive data stored in SaaS applications is vulnerable to leaks, especially when files are shared externally or when employees inadvertently expose data to unauthorized parties. Without adequate visibility, it's challenging to track data flow and prevent leakage.

2. **Shadow IT:** Shadow IT arises when employees adopt SaaS applications without IT oversight, creating security blind spots. Unauthorized SaaS applications often lack proper security controls, posing risks for data leakage, regulatory non-compliance, and unauthorized access.

3. **Lack of Standardized Security Controls:** Each SaaS provider implements security controls differently. If CISOs can standardize security controls across applications, consistency can strengthen the organization's overall security posture.

4. **Inadequate Access Controls and Privilege Creep:** SaaS applications frequently lack robust access controls, resulting in users retaining access to data and applications they no longer need. This "privilege creep" increases the potential for unauthorized access and data exposure.

5. **Vendor Lock-In and Dependency Risks:** Over-reliance on a particular SaaS vendor can lead to risks if that vendor experiences a breach, service outage, or discontinuation. Vendor lock-in can also complicate security management,

especially if transitioning to another platform is challenging.

Critical Practices for Managing the SaaS Footprint

To effectively manage SaaS applications and mitigate associated risks, CISOs should implement a series of best practices, including:

1. **Implement SaaS Discovery and Monitoring Tools:** SaaS discovery tools provide visibility into sanctioned and unsanctioned applications across the organization. These tools help identify and monitor shadow IT, providing insights into application usage, data transfers, and access patterns. Some organizations use Cloud Access Security Brokers (CASBs) to achieve visibility and control over SaaS applications, ensuring data security and compliance.

2. **Create a SaaS Governance Framework:** A governance framework provides a structured approach to SaaS management, defining policies and procedures for acquiring, using, and de-provisioning assets.

   Governance guidelines should include:
   - **Approval Processes:** A formal process for vetting and approving SaaS applications before deployment.
   - **Access and Usage Policies:** Rules around who can access SaaS tools, for what purposes, and with what permissions.
   - **Data Classification and Handling Policies:** Guidelines on what data can be stored or processed within specific SaaS applications.

3. **Enforce Identity and Access Management (IAM) Across SaaS Applications:** Strong IAM controls are essential for SaaS security, ensuring users have the appropriate level of access.

   CISOs should:

- **Implement Single Sign-On (SSO):** SSO allows users to access multiple SaaS applications through one login, simplifying access management and reducing the risk of credential compromise.
- **Enforce Multi-Factor Authentication (MFA):** MFA strengthens access security for SaaS applications, especially for users accessing sensitive or high-risk applications.
- **Regularly Review Access Rights:** Conduct periodic access reviews to ensure users retain only the necessary privileges and remove access when it's no longer required.

4. **Monitor SaaS Activity for Anomalies and Threat Detection:** Monitoring SaaS applications for unusual activity is essential for early threat detection.

   This includes:
   - **User Behavior Analytics:** Identifying unusual patterns, such as high volumes of data downloads or logins from distinctive locations, can signal potential security incidents.
   - **Data Loss Prevention (DLP) Policies:** Enforcing DLP policies in SaaS applications helps prevent sensitive data from being shared or leaked outside the organization.

5. **Assess and Manage SaaS Vendor Risk:** Every SaaS vendor introduces a degree of supply chain risk.

   CISOs should conduct regular security assessments of vendors, including:
   - **Security Certifications and Compliance:** Verifying that vendors meet industry standards (e.g., SOC 2, ISO 27001) and comply with relevant regulations.
   - **Data Protection Measures:** Ensuring vendors implement robust encryption, access controls, and other measures to protect customer data.

- **Incident Response and Recovery Plans:** Review vendors' incident response and business continuity plans to ensure they align with the organization's requirements.

6. **Establish an Incident Response Plan for SaaS Applications:** An incident response plan tailored to SaaS applications enables rapid response to potential security incidents. CISOs should ensure that incident response procedures:

   - **Include Clear Roles and Responsibilities:** Define roles for key stakeholders, including IT, security, and legal teams.
   - **Outline Communication Protocols:** Determine how and when incidents will be communicated to affected parties, including employees, customers, and regulators.
   - **Specify Containment and Recovery Actions:** Identify actions to contain and remediate breaches, such as revoking access or limiting data sharing until the issue is resolved.

Case Studies Illustrating the Importance of SaaS Visibility

1. **Uber's Data Breach through Third-Party SaaS Application:** In 2016, a data breach at Uber exposed sensitive customer and driver information. The breach occurred because attackers gained unauthorized access through compromised credentials for a third-party SaaS platform. This incident underscores the importance of robust access controls and third-party risk management for SaaS applications.

2. **Capital One and Cloud Misconfigurations:** Capital One's 2019 data breach was facilitated by a misconfigured firewall on a cloud-based SaaS application, allowing an unauthorized party to access over 100 million customer records. This case highlights the need for continuous monitoring, configuration management, and security assessments of SaaS applications.

Benefits of a Well-Managed SaaS Footprint

When CISOs invest in understanding and managing their SaaS footprint, the organization benefits in several ways:

1. **Enhanced Security and Data Protection:** Comprehensive visibility into SaaS applications allows CISOs to implement consistent security measures, reducing the risk of data leaks, breaches, and unauthorized access.

2. **Improved Compliance Posture:** By monitoring SaaS applications for compliance with data protection regulations, CISOs can minimize regulatory risks and avoid costly fines.

3. **Cost Savings and Resource Optimization:** Identifying and de-provisioning unused or redundant SaaS applications prevents resource waste, enabling a more cost-effective IT environment.

Chapter 2

KNOW THY BUSINESS

The Importance of CISOs Knowing the Intricacies of their Business Know Thy Business - The Business Imperative

One week after accepting the role of System CISO of the Federal Reserve's National IT Organization in February 2016, the Central Bank of Bangladesh was hit with what has been recorded as the most significant cyber heist in history. Hackers, attributed to being the Lazarus Group and affiliated with the Democratic Republic of North Korea, attempted to steal nearly $1 billion from the Bangladesh Central Bank's account at the Federal Reserve Bank of New York (one of the twelve regional Federal Reserve Banks within my purview).

The hackers used malware to infiltrate the Bank's computer systems, allowing them to access the Bank's network. Once inside, the hackers used fraudulent SWIFT (Society for Worldwide Interbank Financial Telecommunication) messages to request funds transfer from the Bangladesh Bank's account at the Federal Reserve Bank of New York to accounts in the Philippines and Sri Lanka.

The hackers issued thirty-five fraudulent instructions via the SWIFT interbank payment network. Five of the thirty-five fraudulent instructions successfully transferred US$101 million, with US$81 million traced to the Philippines and US$20 million to Sri Lanka. The Federal Reserve Bank of New York blocked the remaining thirty transactions, amounting to US$850 million, due to suspicions raised by a misspelled instruction. Around US$18 million of the US$81 million transferred to the Philippines has been recovered, and all the money transferred to Sri Lanka has since been recovered.

Upon compromising the Bangladesh Central Bank's computer network, the bad actors burrowed themselves deep into the systems, covering their tracks and observing the business processes associated with interbank transfers, having gained access to the Bank's credentials for payment transfers. They used these credentials to authorize about three dozen requests to the Federal Reserve Bank of New York. Overall, the incident highlighted vulnerabilities in the global financial system and led to far more stringent controls embedded in the worldwide banking system. The incident also revealed to me personally, as a practitioner, the importance of "Knowing Thy Business."

Know Thy Business Processes

The extent to which the Lazarus Group took the time to learn the intricacies of the business processes, embed themselves as deeply as they did, and covered their tracks was quite remarkable.

Chief Information Security Officers (CISOs) have assumed a pivotal role in protecting organizations from digital threats in the ever-evolving cybersecurity landscape. While CISOs' technical understanding and cybersecurity expertise are undeniably critical, a deep understanding of the businesses they serve has become equally essential. CISOs, therefore, need to ensure that their cybersecurity strategies align with the business objectives of the organizations they support.

Cybersecurity is not an isolated function but an integral part of the broader business strategy. A CISO who comprehends the business goals, revenue streams, and market dynamics can effectively align cybersecurity initiatives with these objectives. This alignment ensures that security measures are not perceived as impediments to progress but as growth enablers. Furthermore, from a risk management and prioritization perspective, understanding the business intricacies enables CISOs to identify and assess risks specific to the organization. This knowledge empowers them to prioritize security efforts based on the potential impact on critical processes and assets.

By quantifying risks in terms of business consequences, CISOs can engage executive stakeholders in discussions more effectively. CISOs often need to convey complex cybersecurity concepts to non-technical stakeholders. A deep comprehension of the business allows them to translate technical jargon into terms that resonate with board members, executives, and employees across departments. This ensures that cybersecurity matters are understood, appreciated, and integrated into decision-making processes. Generic cybersecurity approaches seldom yield optimal results. CISOs who know the business can tailor security solutions to the organization's needs, technologies, and workflows. This targeted approach enhances the efficiency and effectiveness of security measures while minimizing disruptions. A security-conscious culture requires top-down support and endorsement. CISOs who understand the business can craft security awareness programs that relate the best cybersecurity practices to everyday work scenarios. This approach reinforces the idea that security is everyone's responsibility.

Securing an organization requires financial resources and a workforce. A CISO who comprehends the business can make informed budget requests by highlighting how investments in cybersecurity directly contribute to safeguarding revenue, reputation, and customer trust. Lastly, cybersecurity threats continually evolve, demanding proactive strategies and innovative approaches. A CISO immersed in the business landscape can contribute to strategic planning by anticipating security needs based on upcoming business initiatives and industry trends.

Know Thy Enterprise Architecture

While understanding the business objectives is critical for a CISO, having a firm grasp of the enterprise processes and technical architecture that support the organization's operations is equally important. Together, these three pillars—business, processes, and technology—form the foundation upon which a robust cybersecurity strategy can be built.

Enterprise Business Processes: CISOs must have a deep understanding of how business processes function across departments. These automated or data-driven processes are essential for generating revenue, delivering products, managing customers, or ensuring compliance. Cyber incident disruption of these processes can weaken the entire organization.

Mapping Critical Workflows: CISOs need to work closely with key departmental stakeholders to map out the workflows that are crucial to the business. This could involve understanding the order-to-cash process in a retail company or how patient records are managed in a hospital. Knowing these workflows helps the CISO identify the key points where security risks could cause major operational disruptions.

Business Process Resilience: Beyond identifying risks, the CISO must also work to ensure that business processes are resilient. This could involve introducing backup systems, enforcing tighter access controls, or deploying technologies that allow faster recovery from cyber incidents.

Know Thy Technical Architecture: A CISO's understanding of the business must be complemented by an in-depth knowledge of the organization's technical architecture. The network systems, applications, and databases that support the business's processes must be adequately secured. Without this understanding, even the most well-designed cybersecurity strategy can fall short.

Infrastructure Awareness: Knowing the full extent of the organization's infrastructure is essential. This includes understanding where critical data is stored, how applications interact with one another, and which third-party integrations or cloud services are in use. For example, a CISO at a financial institution needs to know how payment systems are interconnected with databases and third-party services to ensure no vulnerabilities are overlooked.

Emerging Technology and Security: As organizations adopt technologies like AI, IoT, and cloud-based services, the technical architecture becomes even more complex. A CISO must stay informed on the latest security practices related to these technologies and ensure they are appropriately integrated into the organization's infrastructure.

System Interdependencies: CISOs should closely examine how different systems within the organization are interdependent. A disruption in one area (e.g., network outages or ransomware affecting a database) could cascade into a more significant crisis. By understanding these dependencies, the CISO can better plan for mitigating and responding to incidents.

A Tale Of "Business" Email Compromise

The urgency of "Knowing Thy Business" also factored greatly in my role as the then CISO for the Americas of Mitsubishi Union Financial Group (MUFG) Union Bank when Business Email Compromise (BEC) attacks were at an all-time high. MUFG Union Bank is a subsidiary of Mitsubishi UFJ Financial Group (MUFG), one of the world's largest and most diversified financial groups with over $2.3T in assets under management. Union Bank provides various financial services, including banking, wealth management, and commercial banking. US banks have since acquired its retail banking services, and many of the BEC scams I had to defend against during my tenure were traditionally targeted.

By 2022, the Federal Bureau of Investigation (FBI) Internet Crime Complaint Center (IC3) received 21,832 BEC complaints nationwide, with adjusted losses of over $2.7 billion to U.S. businesses. BEC is a sophisticated scam targeting businesses and individuals who are performing fund transfers. The scam is frequently carried out when a subject compromises legitimate business email accounts through social engineering or computer intrusion techniques to conduct unauthorized funds transfers. As fraudsters have become more sophisticated and preventative measures have been put in place, the BEC scheme has continually evolved in kind. The scheme has evolved from simple hacking or spoofing of business and personal email accounts and a request to send wire payments to fraudulent bank accounts. These schemes historically involved compromised vendor emails, requests for W-2 information, targeting of the real estate sector, and fraudulent requests for large amounts of gift cards. More recently, fraudsters are more frequently utilizing custodial accounts held at financial institutions for cryptocurrency exchanges or having victims send funds directly to cryptocurrency platforms where funds are quickly dispersed. In 2022, the IC3 also saw a slight increase in targeting victims' investment accounts instead of traditional banking accounts. There was also an increasingly prevalent tactic by BEC bad actors, who spoofed legitimate business phone numbers to confirm fraudulent banking details with victims.

As the America's CISO of MUFG Union Bank, I led the cross-bank initiative to reduce the Bank's exposure to BEC scams through several related initiatives, enforcing Multi-Factor Authentication (MFA) across the organization, especially for roles like senior executives, financial approvers, systems/network/database administrators, and human resources personnel. I also drove the adoption of advanced email security solutions that included:

- Global threat intelligence with real-time updates for proactive defense against emerging threats
- Forged email detection that blocks customized attacks exploiting executive accounts
- Advanced spam and phishing protection with a high accuracy rate in filtering harmful emails
- URL scanning and filtering to protect users from malicious links
- Domain-based message authentication, reporting, and conformance (DMARC) automated email authentication, which shields your company's email domains from impersonation and potential abuse
- Dynamic malware defense tools with continuous threat analysis, file sandboxing, and automatic breach remediation
- Data Loss Prevention (DLP), complying with regulations to safeguard sensitive outbound information
- User-behavior training modules with threat simulations and education on best practices

I also led a robust security awareness campaign for all employees with continuous security education for frontline banking employees. BEC attackers exploit busy routines, relying on employees overlooking deceptive emails during their busy workday.

Employees were trained to look for these signs of a business email compromise scheme:

- Deadlines are emailed at short notice that involve sending money or sensitive data
- Unusual purchase requests, even when they come from senior officials or trusted colleagues
- Emails from employees sharing new direct deposit details
- Requests to keep information confidential or bypass normal communication channels
- Requests for wire transfers that must be completed hastily or without proper authorization

- Misspellings, grammar, or language that is unusual for the sender
- Something that feels "off" or doesn't look right

I also drove the adoption of the 'four eyes' principle for transactions exceeding specific dollar thresholds and implemented procedures to verify payments and purchase requests outside of e-mail communication. This can include direct phone calls to a known verified number and does not rely on information or phone numbers included in the e-mail communication.

In conclusion, the modern CISO is not solely a guardian of digital defenses; they are strategic partners safeguarding the organization's present and future. A deep understanding of the business and enterprise architecture, positions CISOs to elevate , influence, and impact their role. By fusing technical expertise with business acumen, CISOs become architects of resilience, driving security and growth hand in hand.

Chapter 3

KNOW THY INDUSTRY

Know Thy Industry – The Significance of Industry Acumen for CISOs

In the rapidly evolving realm of cybersecurity, CISOs hold the mantle of safeguarding organizations against an ever-expanding array of digital threats. While technical expertise is undeniably crucial, a profound understanding of the industry's dynamics has become equally indispensable. Being a Chief Information Security Officer (CISO) for a Bank comes with a very different set of expectations than being a Chief Information Security Officer (CISO) for one of the world's largest accounting firms.

The fundamental 'blocking and tackling' remains much the same; however, the dynamics of the operating environments are dramatically different, and the recognition and realization of those subtleties are important for CISOs operating in those environments but critical for CISOs transitioning into those environments from non-adjacent industries.

When I transitioned from leading the cyber program for the U.S. Central Banking System in September of 2019 to joining one of the world's largest professional services firms globally, providing audit, tax, and advisory services to clients in various industries, my mental models had to change. The North American division of Klynveld, Peat, Marwick, and Goerdeler, known by its popular acronym (KPMG), offers a wide range of services to businesses, governments, and non-profit organizations. These services include audit and assurance, tax planning and compliance, advisory services (such as management consulting, risk consulting, and deal advisory), and specialized services like forensic accounting and sustainability consulting.

Quite differently from the Federal Reserve, KPMG operates in over 150 countries and territories, making it one of the largest professional services firms in the world. It has a network of member firms that are legally separate entities but operate under the KPMG brand and share resources, methodologies, and best practices. Another notable difference is that KPMG serves clients across various industries, including financial services, technology, healthcare, manufacturing, energy, and more. It tailors services to each industry's needs and challenges, leveraging industry-specific knowledge and expertise.

As you can imagine, the array of threats and types of threat actors targeting a central banking system, notably the Central Bank of the world's foremost democracy, dramatically differs from those targeting a firm like KPMG. Of course, the same can be said about the regulatory landscape within which both institutions operate.

My transition from KPMG to MUFG Union Bank, North America, heralded my return to the familiar space of the financial services sector after having spent over a decade at the IRS and half a decade at Automatic Data Processing (ADP) Inc., with three years at the Federal Reserve. Whereas the financial motivations of threat actors targeting the IRS, the Federal Reserve, ADP, or MUFG are similar, distinct differences warrant further dissection.

Each industry I've worked in has presented its dynamics from a threat landscape perspective shaped by specific technologies, regulations, and market trends. A CISO with industry acumen must understand the threats and actors most likely to target their sector and particular organizations. This knowledge forms the bedrock of proactive defense strategies tailored to preempt these sector-specific and company-targeted attacks.

In the early 2010s, threat actors worldwide were busy perfecting their attack patterns to target unsuspecting Internet users. Back then, ADP was one of the topmost phished brands in the U.S. Interestingly enough, the same was true for the IRS.

The FBI's Internet Crime Complaint Center (IC3) reported that Business Email Compromise (BEC) scams resulted in over $1.8 billion in losses in that period. This cybercrime involves manipulating employees into transferring funds or sensitive information through deceptive emails.

Back then, the popular anti-phishing protocols and solutions were in their nascent stages of development and adoption: Senders Policy Framework (SPF), Domain Key DKIM, and Domain Message Authentication Reporting and Conformance. ADP was among the very early adopters of these protocols.

SPF serves as an email authentication protocol that helps prevent email spoofing. It allows domain owners to specify which mail servers are authorized to send emails on behalf of their domain. When an email is received, the recipient's mail server can check the SPF record of the sender's domain to verify if the sending mail server is allowed to send emails to that domain. DKIM (Domain Keys Identified Mail) is another email authentication method that helps detect email spoofing and ensures message integrity. It works by adding a digital signature to the email header using cryptographic techniques. The receiving mail server can verify this signature using the public key published in the sender's DNS records. If the signature is valid, it confirms that the email was not altered during transit and originated from the claimed sender. DMARC (Domain-based Message Authentication, Reporting, and Conformance) builds upon SPF and DKIM to provide an additional layer of email authentication and reporting. It allows domain owners to specify how their emails should be handled if they fail SPF or DKIM checks. Additionally, DMARC enables domain owners to receive reports from email receivers about the disposition of emails claiming to be from their domain (e.g., delivered, quarantined, or rejected). This feedback loop helps domain owners monitor and improve email authentication practices.

By proactively adopting SPF, DKIM, and DMARC at both ADP and MUFG Union Bank, I was able to help these organizations reduce their exposure and, most importantly, the exposure of their customers to the rampant phishing attacks that were plaguing those organizations at the time.

It is important to note that industries are bound by distinct regulatory frameworks that mandate cybersecurity practices. CISOs versed in their industry dynamics can ensure compliance by aligning security protocols with these regulations. This knowledge not only averts potential penalties but also cultivates a culture of trust with clients and partners.

Within the U.S. banking and finance sector, it is reported that CISOs allocate, on average, approximately 40% or more of their time/attention and budgets to address regulatory compliance matters. I found that painfully true during my tenure at MUFG Union Bank, regulated by the Federal Reserve, the Office of the Comptroller of the Currency (OCC), and the Federal Deposit Insurance Corp (FDIC).

Industry dynamics often foster collaboration among peers facing similar security challenges. A CISO who comprehends these dynamics can leverage industry networks to share threat intelligence, strategies, and best practices. This collaborative approach bolsters cyber defenses across the industry ecosystem, especially in heavily regulated environments, which arguably poses both a challenge and a massive opportunity for CISOs. It requires a strategic approach to compliance, where understanding and anticipating regulatory requirements become integral to cybersecurity planning. Compliance should not be seen merely as a checklist but as a fundamental component of the organization's security posture. At the same time, integrating regulatory requirements into cybersecurity frameworks can help CISOs ensure that their organizations meet the minimum compliance standards and protect sensitive information.

The strategic importance of regulatory knowledge for CISOs cannot be overstated. In an era where data breaches can lead to significant financial penalties and irreparable damage to reputation, ensuring compliance with relevant regulations is critical to a comprehensive cybersecurity strategy. CISOs, therefore, must go beyond their traditional roles to become champions of compliance, integrating legal and regulatory requirements into the fabric of their cybersecurity initiatives. Aligning cybersecurity policies with regulatory requirements is a complex but necessary endeavor. Regulations like the Global Data Protection Regulation (GDPR), the Health Insurance Portability and Accountability Act (HIPAA), and the Payment Card Industry Data Security Standard (PCI DSS) dictate stringent standards for data protection, requiring organizations to implement specific security measures. CISOs must ensure that their cybersecurity frameworks meet these standards and are flexible enough to adapt to new regulations as they emerge. This requires a deep understanding of both the letter and the spirit of the rules, enabling CISOs to anticipate changes and adjust their strategies accordingly.

In light of these, adopting a proactive approach to regulatory standards is crucial for CISOs. This involves:

- **Continuous Monitoring:** Keeping abreast of regulatory developments and interpretations may affect the organization's compliance status.
- **Engagement and Advocacy:** Participating in discussions and forums on cybersecurity and regulatory issues, allowing CISOs to influence policy and understand emerging trends.

- **Education and Training:** Ensuring the cybersecurity team and broader organization know compliance requirements and their roles in maintaining them.

For CISOs, integrating regulatory knowledge into cybersecurity strategies is not just a compliance exercise but a strategic move to protect the organization and its stakeholders. This integration involves a multi-faceted approach, encompassing the adaptation of cybersecurity policies, continuous monitoring of regulatory changes, and fostering a culture of compliance within the organization.

Common Challenges in Regulatory Compliance include:

- **Rapidly Evolving Regulations:** The pace at which regulatory frameworks are updated can be daunting. Keeping abreast of new regulations and amendments, especially for international organizations, requires constant vigilance.
- **Complexity and Diversity of Regulations:** The sheer volume and complexity of regulations such as the Global Data Protection Regulation (GDPR), Personal Information Protection & Electronic Document Act (PIPEDA), Federal Financial Institutions Examination Council (FFIEC), Health Insurance Portability and Accountability Act (HIPAA), and Payment Card Industry (PCI) Data Security Standard (DSS), each with its requirements, create a challenging environment for CISOs to ensure comprehensive compliance.
- **Integration of Compliance into Existing Systems:** Retrofitting compliance measures into established systems and processes can be technically challenging and resource-intensive.
- **Cross-Functional Collaboration:** Ensuring effective communication and collaboration between cybersecurity, legal, compliance, and business units is often easier said than done, yet it's critical for cohesive regulatory compliance.

Staying informed about regulatory changes and interpretations is crucial for CISOs. This requires a proactive strategy that includes:

- **Regulatory Tracking Tools:** Leveraging software and online platforms that provide updates on regulatory changes, helping CISOs stay ahead of new requirements and amendments.
- **Professional Networks and Forums:** Participating in industry groups, forums, and professional networks offers insights into how peers address regulatory challenges and provides a platform for sharing best practices and solutions.
- **Ongoing Education:** Engaging in continuous education through seminars, workshops, and courses on regulatory compliance to understand the nuances of various legal frameworks and their implications for cybersecurity.

The heart of integrating regulatory knowledge into cybersecurity strategies lies in embedding these requirements into the organization's cybersecurity framework. This involves:

- **Policy Development and Revision:** Regularly updating cybersecurity policies to reflect current regulatory requirements, ensuring all measures are in place to ensure compliance.
- **Risk Assessment and Management:** Conduct thorough assessments of compliance risks. This involves identifying areas where the organization's cybersecurity practices may not meet regulatory standards and developing action plans to address these gaps.
- **Cross-departmental collaboration:** Collaborating with legal, compliance, and operational teams to ensure a holistic approach to regulatory compliance. This synergy ensures that cybersecurity strategies are not developed in isolation but are informed by comprehensive organizational knowledge and perspectives.

While integrating regulatory knowledge into cybersecurity strategies is critical, it presents challenges, including the complexity of regulations and the rapid pace of change. Best practices to address these challenges include:

- **Leverage Technology for Compliance Management:** Advanced compliance management tools can automate the tracking of regulatory changes, assess compliance levels, and manage documentation. These tools can significantly reduce the manual workload and increase accuracy in compliance reporting.
- **Develop a Regulatory Change Management Process:** Establish a formal process for identifying, assessing, and implementing responses to regulatory changes. This should include mechanisms for rapidly disseminating information about regulatory updates to all relevant stakeholders.
- **Foster a Culture of Compliance:** Embed compliance into the corporate culture by making it a shared responsibility across all levels of the organization. Regular training and awareness programs help ensure that every employee understands their role in maintaining compliance.
- **Collaborate Across Departments:** Create cross-functional teams that include members from legal, compliance, IT, and business units to ensure a holistic approach to regulatory compliance. These teams can provide diverse perspectives and facilitate the integration of compliance requirements into all aspects of business operations.
- **Engage with Regulatory Bodies and Industry Groups:** Active engagement with regulatory bodies and participation in industry groups can provide insights into regulatory trends and best practices. It also offers opportunities to influence policy development and gain early visibility into upcoming regulatory changes.
- **Utilize External Expertise:** Consulting with external legal and compliance experts can provide valuable guidance and

help avoid missteps in complex regulatory environments or when entering new markets.

For CISOs, particularly in the financial sector, managing relations with regulators such as the Federal Reserve Bank examiners and the Office of the Comptroller of the Currency (OCC) bank examiners is another crucial aspect of the compliance process. This relationship is about adhering to regulations and fostering an environment of transparency, cooperation, and mutual understanding.

Banking regulators ensure the financial system's stability, integrity, and security. From their perspective, cybersecurity is not just about protecting individual institutions but also about safeguarding the broader economic ecosystem from systemic risks. CISOs must recognize this broader mandate to thoroughly appreciate the motivations behind regulatory examinations and requirements.

Establishing Open Lines of Communication

- **Proactive Engagement:** Initiate dialogue with regulators before formal examinations and audits. This proactive approach can help establish a foundation of trust and cooperation.
- **Transparency:** Be transparent about your organization's cybersecurity strategies, challenges, and risk management practices. This openness can foster a more constructive and less adversarial relationship.
- **Regular Updates:** Keep regulators informed about significant changes in your cybersecurity posture or new initiatives undertaken. Regular updates can demonstrate your commitment to compliance and continuous improvement.

Leveraging Regulatory Interactions for Insights

- **Feedback Loops:** Treat feedback from regulators not as a compliance checklist but as valuable insights into potential vulnerabilities or areas for improvement.
- **Best Practices Sharing:** Regulators often have a broad view of the industry and can share anonymized best practices or trends they observe across other institutions. Engaging in such exchanges can provide CISOs with valuable benchmarks and innovative strategies.

Navigating Challenges and Disagreements

- **Constructive Disputes:** When disagreements arise, approach them constructively, presenting data or evidence to support your position. It's essential to frame disputes as professional differences in interpretation rather than confrontational stand-offs.
- **Seeking Clarification:** Regulatory language can sometimes be ambiguous. When in doubt, seek clarification directly from regulators to ensure your organization's practices align with the intended regulatory requirements.

Effectively managing relations with banking regulators is critical to the CISO's role, requiring a delicate balance between compliance, communication, and collaboration. By understanding the regulators' perspective, establishing open lines of communication, leveraging regulatory interactions for insights, navigating challenges constructively, and building a culture of regulatory engagement, CISOs can enhance their organizations' compliance posture and contribute to the broader goal of financial systems integrity.

Chapter 4

KNOW THY SOFTWARE

Knowing Thy Hardware Assets is critical, but equally important is "Knowing Thy Software Assets

Software Asset Management (SAM) is a strategic approach to managing an organization's software assets throughout their lifecycle. Although traditionally focused on the processes, policies, and procedures that enable IT managers to gain control over their software licenses, installations, and usage. The importance of SAM for cyber defenders cannot be understated.

In the 'thick of the fight,' having access to robust SAM processes affords cyber defenders' critical insights to formulate battle damage assessment or impact analysis while ensuring enterprise software is kept up to date with the latest security patches, which is crucial for protecting against vulnerabilities and cyber threats. SAM helps track software versions and apply patches promptly, reducing the risk of security breaches. Malicious actors can exploit unused or outdated software, making SAM a critical component of an organization's cybersecurity strategy. Separately and from a third-party risk management perspective, SAM empowers organizations during vendor negotiations, especially regarding security terms and conditions.

A clear understanding of software security, usage, and license requirements gives businesses an advantage when discussing contracts and terms with software vendors. This can lead to more favorable agreements and pricing structures, ultimately saving costs and improving security posture in the long run. Although tangentially related to the practice of cybersecurity, SAM contributes to business continuity by ensuring organizations have access to essential software applications during unforeseen events or disasters, such as the effects of a debilitating malware attack. Companies can quickly restore critical software systems by having a well-documented inventory of software licenses and installation files, minimizing downtime and ensuring smooth operations during challenging situations.

Directly related to the necessity of having robust Software Asset Management in place, a closely related practice area that has recently grown in urgency and importance is the need for a reliable Software Bill of Materials (SBOM). SBOM is a comprehensive list of components and ingredients used to build software. It provides detailed information about open-source and third-party components, libraries, frameworks, and dependencies in a software application.

Reasons why CISOs should prioritize SBOM as a part of their SAM strategy include the following:

- **Supply Chain Security:** SBOM enhances the transparency of the software supply chain. With cyber threats becoming increasingly sophisticated, cyber defenders must ensure the security and integrity of the software their organization's critical business applications rely upon. SBOM allows organizations to identify and track components, making detecting vulnerabilities and potential security risks in the supply chain easier. By knowing what elements are used, cyber defenders can quickly respond to security incidents, apply patches, and prevent potential breaches.

- **Rapid Vulnerability Identification and Patching:** SBOM enables organizations to swiftly identify software vulnerabilities by providing a detailed inventory of components. This proactive approach allows cyber defenders to quickly match known vulnerabilities to their specific software components. Once vulnerabilities are identified, organizations can promptly apply patches or updates, reducing the exposure window and minimizing the risk of exploitation by cybercriminals.

- **Compliance and Licensing Management:** Open-source Software often comes with specific licenses that organizations must adhere to. SBOM helps accurately identify and track open-source components and their respective licenses. This ensures that organizations comply with licensing agreements and avoid legal issues related to intellectual property rights. Proper licensing management also helps organizations make informed decisions about their software, considering costs, licensing restrictions, and compatibility.

- **Risk Mitigation and Decision-Making:** Understanding software composition through SBOM allows organizations to assess the risk associated with different components. It enables informed decision-making, helping organizations choose components with better security records and reliability. Organizations can build more secure and resilient software applications by mitigating risks associated with vulnerable or outdated components.

In summary, the Software Bill of Materials (SBOM) is crucial in enhancing supply chain security, enabling rapid vulnerability identification, ensuring compliance with licenses, mitigating risks, supporting decision-making, promoting standardization, and aiding regulatory compliance. By implementing SBOM practices, cyber defenders can significantly improve their software security posture and minimize the potential of cyber threats and vulnerabilities.

Know Thy Data

The Crucial Role of Data Inventory Management for CISOs

In today's digital landscape, data inventory management has become a cornerstone of effective cybersecurity and risk management for CISOs. With regulatory mandates tightening globally, data exponentially growing, and cyber threats becoming more sophisticated, managing data inventory is more than just an operational necessity—it's a strategic imperative.

This chapter delves into why data inventory management is essential for CISOs, covering the benefits, challenges, regulatory pressures, and best practices that help CISOs safeguard their organizations.

1. **Understanding Data Inventory Management:** Data inventory management is the systematic cataloging and organizing of all data assets. This involves creating a precise inventory of all data types, formats, storage locations, owners, and associated risk levels. This inventory often extends to tracking data flow within and outside the organization to manage data in transit effectively.

   Key components of data inventory management include:
   - **Data Identification:** Identifying what data exists, including structured and unstructured data.
   - **Data Classification:** Categorizing data based on sensitivity, regulatory requirements, and business importance.
   - **Data Mapping:** Documenting where data resides, how it flows, and where it is stored, both on-premises and in the cloud.

- **Ownership and Access Control:** Assigning responsibility for data assets and managing access controls.
2. **The Business Case for Data Inventory Management:**
   A. **Enhancing Data Security and Minimizing Cyber Risks**

   Data inventory management enables a CISO to pinpoint where sensitive data resides, who has access, and how it moves across the organization. Data vulnerabilities remain hidden without this insight, and attackers can exploit these weaknesses.

   - **Risk Identification:** With a thorough data inventory, CISOs can assess the risk profile of each data asset, ensuring that data critical to business operations or customer privacy is prioritized for protection.
   - **Efficient Incident Response:** Knowing where data resides and how it is accessed allows for quicker response times in breach scenarios. It also minimizes downtime and data exfiltration.
   - **Minimizing Attack Surface Data:** By reducing shadow IT and unknown data sources, as well as data inventory management, creates an organization's exposure to attack vectors.

   B. **Compliance and Regulatory Obligations**

   Compliance with data protection laws has become a priority for organizations worldwide. Regulations like GDPR, CCPA, HIPAA, and SOX impose stringent requirements for handling, storing, and protecting data. Data inventory management is crucial for demonstrating compliance and avoiding costly fines.

   - **Regulatory Alignment:** Regulations require organizations to understand and monitor how personal and sensitive data is used. A well-maintained data inventory ensures that data handling aligns with these legal requirements.

- **Audits and Reporting:** With an updated data inventory, CISOs can quickly provide necessary documentation during audits, avoiding penalties for non-compliance.
- **Data Subject Rights:** Under laws like GDPR, users have rights such as the right to access, rectify, and delete personal data. Data inventory management enables CISOs to locate and respond to user data requests efficiently.

C. Data Governance and Policy Enforcement

Data inventory management is foundational for developing a robust data governance framework. Without a clear view of data assets, CISOs cannot enforce consistent policies on data handling, retention, and protection.

- **Consistent Policy Application:** Policies for data access, handling, and retention can be uniformly applied, reducing the likelihood of mishandling sensitive information.
- **Transparency and Accountability:** With a well-maintained data inventory, CISOs can assign data ownership, hold departments accountable for data practices, and ensure adherence to data protection policies.
- **Improved Decision Making:** Inventory management provides CISOs real-time insights into data trends and usage patterns, aiding data-driven decisions on security investments and resource allocation.

D. Cost Efficiency

Poor data management leads to inefficiencies, including duplicated data storage, increased data processing costs, and extended response times. By managing data inventories effectively, CISOs can eliminate redundant data, optimize storage, and reduce expenses.

- **Reducing Storage Costs:** Identifying and eliminating duplicate or redundant data reduces the storage footprint and associated costs.
- **Improving Efficiency:** With a clear understanding of data assets, security teams can prioritize their focus on high-value data and assets, reducing wasted effort in low-impact areas.
- **Enhanced ROI on Security Investments:** Knowing where sensitive data is and how it's used ensures that CISOs can allocate security resources—like DLP, encryption, or segmentation—most effectively.

3. **Challenges in Data Inventory Management:**
   A. **Complexity of Modern Data Environments:** With data stored across on-premises, cloud platforms, and third-party applications, maintaining a comprehensive inventory is challenging. Data flows are dynamic, and new data sources frequently emerge, making inventory management a continuous effort.
   B. **Data Silos:** Data often exists in isolated silos across departments, making it difficult for CISOs to get a centralized view. Overcoming this requires collaboration with IT and business units to break down these silos.
   C. **Evolving Regulatory Landscape:** With data privacy laws constantly changing, CISOs must frequently update data inventory practices to stay compliant. Failing to adapt to new regulations or requirements can result in non-compliance and potential penalties.
   D. **Shadow IT and Unstructured Data:** Shadow IT (technology used without explicit organizational approval) complicates data management efforts. Additionally, un-

structured data (emails, documents, images) is more complex to catalog, requiring advanced technologies like AI and machine learning for inventory and analysis.

Best Practices for Effective Data Inventory Management

- **Establish Clear Ownership and Accountability:** Assign data owners responsible for cataloging and maintaining data. This not only aids inventory management but also ensures departments are held accountable for their data security practices.
- **Use Automated Data Discovery Tools:** With tools like data discovery and classification software, CISOs can automate inventory tasks, continuously identifying and cataloging new data. Automation is essential to keep pace with the evolving data landscape.
- **Implement Regular Audits and Updates:** Data inventories need regular audits to remain accurate. Establish periodic reviews to remove outdated records, assess compliance, and identify potential risks.
- **Leverage Data Classification:** Classify data by sensitivity (e.g., public, confidential, restricted) to streamline protection efforts. This helps CISOs apply appropriate security controls without over burdening low-risk data.
- **Integrate with Broader Cybersecurity Initiatives:** Data inventory should be integrated with other security initiatives, such as incident response, threat hunting, and vulnerability management. Cross-functional collaboration ensures that inventory management supports broader cybersecurity goals.

Emerging Technologies and Trends in Data Inventory Management

- **A.I. and Machine Learning:** AI-powered tools can assist in cataloging and analyzing data flows automatically. They can also detect anomalies and identify data that may not be inventoried appropriately, reducing the manual workload.
- **Privacy Management Platforms:** Privacy platforms are increasingly used to track data subject rights, automate compliance processes, and manage data inventories in line with regulatory requirements.
- **Zero Trust Architecture:** Zero trust principles rely on strict access controls based on identity, device, and data location. An adequate data inventory is fundamental to implementing zero trust, ensuring only the right individuals can access specific data assets.

In an era where data drives business decisions, and cyber threats evolve rapidly, data inventory management has become essential for CISOs to maintain a secure, compliant, and efficient organization. This practice enables security leaders to identify and mitigate risks, ensure compliance with complex regulatory landscapes, enforce data governance policies, and optimize operational costs. As data grows in volume, velocity, and variety, CISOs must prioritize data inventory management to safeguard organizational assets, maintain regulatory compliance, and support the broader goals of risk management and business resilience.

Chapter 5

KNOW THY CLOUDS

Know Thy Clouds - Why CISOs Must Understand Their Cloud Security Status

In December 2018, after two years as CISO for the U.S. Central Banking System, and while the Federal Reserve prepared to introduce public clouds to drive innovation and transformation, reports emerged that Chinese hackers broke into Hewlett Packard and IBM networks, then used the access to hack into their clients' computers, according to five sources familiar with the attack. The attacks were part of a Chinese campaign called Cloud Hopper, which the United States and Britain said infected technology service providers to steal secrets from their clients. Cloud Hopper targeted MSPs to access client networks and steal corporate secrets from companies around the globe, according to a U.S. federal indictment of two Chinese nationals.

It goes without saying that Cloud Hopper was a 'watershed' cloud security breach with many key lessons and broad implications not just for me and CISOs industry-wide, but also for Third-Party Risk Management (TPRM) programs in general.

Key Points of the Breach

Security weaknesses in cloud configurations and access controls led to unauthorized access. These vulnerabilities were reportedly due to misconfigurations and inadequate monitoring within the HPE cloud services infrastructure.

Data Exposure: Sensitive information, including Personally Identifiable Information (PII), financial records, and client data, was exposed. Due to HPE's extensive client base and DXC's acquisition of HPE's services, the breach had a broad impact, affecting customers across sectors, including finance, healthcare, and government.

Implications for CISOs in Third-Party Risk Management

The HPE/DXC cloud breach highlighted several critical areas for CISOs, particularly cloud security and vendor oversight.

- **Thorough Cloud Vendor Due Diligence:** When partnering with cloud providers, CISOs must conduct rigorous evaluations of security practices, focusing not only on certifications (such as System and Organization Controls "SOC" 2 or ISO 27001) but also on the provider's ability to prevent misconfigurations. Review the vendor's architecture, access controls, and ongoing security practices to ensure they meet organizational standards.
- **Cloud Configuration and Access Control Audits:** The breach illustrates the importance of monitoring cloud configurations continuously. CISOs should implement regular audits of all cloud configurations and apply least-privilege access policies to prevent unauthorized access.
- **Vendor Service Level Agreements (SLAs) and Security Requirements:** Contracts with cloud vendors should in-

clude detailed SLAs that establish requirements for security practices, incident response, and regular vulnerability assessments. SLAs should require immediate notification of security incidents affecting customer data.

- **Data Segmentation and Multi-Tenancy Risk:** Organizations must verify that cloud vendors segment data securely and apply adequate security to multi-tenant environments. Segmentation helps limit data exposure in a breach within a shared cloud infrastructure.

- **Collaborative Incident Response Plans:** Establish well-defined, collaborative incident response protocols with cloud vendors, ensuring that any detected incident triggers immediate action from both parties. Effective collaboration reduces incident impact and enhances response effectiveness.

As organizations continue to migrate their Infrastructure and services to the cloud, the need for robust cloud security has become paramount. Cloud computing offers unparalleled scalability, flexibility, and cost efficiency, empowering organizations to innovate and grow faster. However, with these benefits come new and unique security challenges that CISOs must navigate. The traditional security perimeter no longer applies, as data, applications, and workloads reside in multiple environments—public cloud, private cloud, hybrid, and multi-cloud setups.

For CISOs, understanding the organization's cloud security status is essential for maintaining control, ensuring compliance, and protecting critical assets. Without a clear, real-time view of the cloud security landscape, organizations face increased risks, from data breaches and misconfigurations to compliance failures and reputational damage. This chapter explores the importance of understanding cloud security status, the challenges associated with cloud environments, and actionable strategies to enhance visibility and control over cloud security.

The Evolution of Cloud Adoption and Security Complexity

In recent years, cloud adoption has skyrocketed. Organizations now rely on cloud platforms for mission-critical functions, including data storage, software development, collaboration, and customer relationship management. However, as companies shift workloads to the cloud, they face a set of security considerations that differ significantly from traditional on-premises environments. In cloud environments, security responsibilities are shared between the organization and the cloud provider, creating a **shared responsibility model** that varies by service type (e.g., Infrastructure as a Service (IaaS), Platform as a Service (PaaS), and Software as a Service (SaaS)).

This shift has introduced the need for CISOs to understand:

- **Security Posture Across Different Cloud Models**: It is important to know which security responsibilities lie with the organization and which are managed by the cloud provider.
- **Visibility Across Multi-Cloud Environments**: Many organizations use multiple cloud providers, which can create silos and complicate monitoring and response.
- **Cloud-Specific Security Controls**: Cloud environments introduce unique risks, such as account hijacking, data exposure, and configuration errors, necessitating cloud-native security solutions.

Why Cloud Security Status is Crucial for CISOs

Understanding cloud security status allows CISOs to secure data and systems in dynamic and distributed environments effectively. Some of the critical reasons for prioritizing cloud security include:

- **Preventing Data Breaches and Unauthorized Access:** Cloud environments store vast amounts of sensitive data. Misconfigurations, weak access controls, or poorly managed keys can lead to significant data breaches.
- **Managing Complex Access and Identity Requirements:** Identity and Access Management (IAM) is critical in cloud security, where identity is often the new perimeter. Mismanaged identities can allow unauthorized access to sensitive data and systems.
- **Ensuring Compliance Across Geographies:** Regulatory compliance is a top priority for organizations, especially those handling sensitive or regulated data. CISOs need a clear view of data flows, encryption practices, and audit trails to ensure compliance in the cloud.
- **Mitigating Configuration Risks:** Misconfigured cloud resources are a top vulnerability in cloud environments. CISOs must ensure that cloud configurations are continuously monitored and alerts for potential risks are acted upon promptly.
- **Strengthening Incident Response in the Cloud:** Effective incident response in a cloud environment requires a deep understanding of the cloud's architecture and native tools. Without this, incident response can be slow, leading to prolonged breaches and more severe damage.

Common Cloud Security Risks and Threats

The cloud environment introduces risks that differ from those in traditional data centers. Some of the primary cloud security risks include:

Data Breaches Due to Misconfiguration: Cloud misconfigurations, such as public-facing storage buckets, open databases, and unsecured API endpoints, are common causes of data breaches. Attackers can exploit these misconfigurations to gain un-authorized access to sensitive data.

Weak Identity and Access Management (IAM): Access con-trol is crucial in the cloud. Weak IAM policies, such as poor pass-word hygiene, lack of Multi-Factor Authentication (MFA), or ex-cessive privileges, can lead to account takeovers and unauthorized access.

Insider Threats: Cloud environments are particularly vulner-able to insider threats, as privileged users or contractors with ac-cess to cloud resources may intentionally or unintentionally expose data or configurations.

Data Loss: Accidental deletion, lack of data redundancy, and limited backup capabilities can lead to data loss in the cloud. Or-ganizations that rely on their cloud provider for backups without implementing their data resilience strategies risk losing critical in-formation.

Insecure APIs: APIs are essential to cloud functionality, en-abling integration and automation. However, insecure APIs can ex-pose data and introduce vulnerabilities if not properly secured.

Limited Visibility and Monitoring: Cloud environments often need more visibility than on-premises infrastructure. Limited in-sight into data flows, network activity, and user behavior can pre-vent early detection of security incidents.

Critical Practices for CISOs to Enhance Cloud Security

To effectively manage cloud security and mitigate risks, CISOs should implement the following best practices:

Implement Cloud Security Posture Management (CSPM) Solutions: CSPM tools continuously monitor cloud environments for security risks, including misconfigurations, compliance viola-tions, and suspicious activity. These tools provide visibility into cloud resources and configurations, enabling CISOs to address vulnerabilities proactively. CSPM solutions are precious in identifying and correcting configuration drift, which occurs when cloud configurations change over time.

Establish Strong Identity and Access Management (IAM)

Effective IAM is essential for securing cloud environments, where identity is often the primary perimeter. CISOs should:

- **Enforce Multi-Factor Authentication (MFA):** MFA adds an extra layer of protection, especially for privileged accounts and access to sensitive data.
- **Implement Role-Based Access Control (RBAC):** Assign permissions based on job roles, ensuring users have the minimal access necessary to perform their functions.
- **Review Access Privileges Regularly:** Conduct periodic reviews to ensure only authorized users retain access and remove no longer needed privileges.

Encrypt Data at Rest and in Transit

Data encryption is essential in protecting sensitive information within cloud environments. CISOs should:

- **Enforce Encryption Policies:** Ensure all sensitive data is encrypted at rest and in transit, with encryption keys managed securely.
- **Implement Key Management Practices:** Securely store and rotate encryption keys using centralized critical management services or cloud-native essential management tools.

[OBJ]

[OBJ]

Deploy Security Information and Event Management (SIEM) for Cloud Monitoring

SIEM solutions help aggregate and analyze security events from multiple sources across cloud and on-premises environments. A cloud-aware SIEM enables:

- **Centralized Monitoring:** Centralize log data from various cloud providers for a holistic view of security events.
- **Anomaly Detection and Alerting:** Detect anomalies and receive alerts on suspicious activity, allowing the security team to respond swiftly.
- **Incident Correlation Across Environments:** Correlate cloud events with on-premises incidents to comprehensively understand potential threats.

Establish Data Loss Prevention (DLP) Policies

DLP tools help monitor and control the movement of sensitive data within and outside the organization. In cloud environments, DLP is critical for:

- **Preventing Data Leakage:** Control unauthorized data transfers to mitigate data exfiltration risks.
- **Monitoring Data in Transit and at Rest:** Enforce policies for data in motion and at rest to prevent accidental exposure or sharing of sensitive data.

Leverage Cloud Provider Security Tools

Major cloud providers (e.g., AWS, Azure, Google Cloud) offer built-in security tools and services, such as AWS GuardDuty, Azure Security Center, and Google Cloud Security Command Center. CISOs should utilize these tools for:

- **Threat Detection and Vulnerability Scanning:** Identify potential threats and vulnerabilities specific to the cloud provider's environment.
- **Compliance Reporting:** Automate compliance reporting to simplify audits and meet regulatory requirements.
- **Configuration Management:** Monitor and enforce security configurations to prevent misconfigurations.

Develop a Cloud-Specific Incident Response Plan

Incident response in the cloud requires specific knowledge of cloud architecture and cloud provider tools. CISOs should:

- **Train Security Teams on Cloud Incident Response:** Ensure teams understand cloud-specific incident response practices, including containment and remediation.
- **Leverage Cloud Provider Support:** Cloud providers offer incident response services and tools to assist with cloud-specific threats.
- **Establish Clear Communication Channels:** Define communication protocols for incidents involving cloud providers, internal stakeholders, and third-party vendors.

Cloud Compliance and the Role of CISOs

Compliance in the cloud can be challenging due to varying data protection laws across regions and the shared responsibility model. Key compliance considerations include:

- **Data Residency and Sovereignty:** Many regulations require data storage in specific geographic locations. CISOs should ensure cloud providers comply with these requirements, mainly when using global data centers.
- **Automated Compliance Monitoring:** Using CSPM and other compliance tools, CISOs can monitor cloud configurations for compliance with standards like GDPR, HIPAA, and CCPA, reducing the risk of regulatory penalties.
- **Audit Trails and Reporting:** Comprehensive logging and monitoring help maintain audit trails necessary for compliance. CISOs should implement centralized logging and reporting for easy retrieval during audits.

Benefits of a Well-Managed Cloud Security Posture

When CISOs prioritize cloud security, the organization benefits through:

- **Reduced Risk of Data Breaches and Incidents:** Comprehensive cloud security reduces the likelihood of data breaches and service interruptions, protecting the organization's assets and reputation.
- **Enhanced Compliance Posture:** Proactively managing cloud security and compliance ensures adherence to regulatory requirements, reducing the risk of fines or legal consequences.
- **Increased Operational Efficiency:** A well-architected cloud security strategy streamlines management, reducing

redundancies and freeing resources for other critical security functions.

- **Improved Incident Response Capabilities:** With an established cloud-specific incident response plan, security teams can respond swiftly to incidents, minimizing damage and recovery time.

Conclusion: Cloud Security as a Strategic Imperative for CISOs

The cloud offers transformative benefits, but it also introduces complex security challenges. For CISOs, understanding their organization's cloud security status is essential for safeguarding data, maintaining compliance, and mitigating risks. By implementing robust cloud security practices, adopting cloud-native security tools, and continuously monitoring the cloud environment, CISOs can protect their organization in this new frontier of cybersecurity.

In the cloud era, where the perimeter is everywhere, CISOs who actively manage and understand their cloud security status protect and position their organization for secure and sustainable growth. The importance of knowing the cloud security landscape cannot be overstated—it is an essential step toward building a resilient, forward-looking security posture in an increasingly digital world.

Chapter 6

KNOW THY IDENTITIES

The Essential Role of Understanding the Identity an Access Management (IAM) Estate: Human and Non-Human Identities

In July 2019, as my tenure as the CISO of the Federal Reserve was winding down, in another watershed incident with industry-wide implications. Capital One disclosed that Paige Thompson, a former Amazon Web Services (AWS) employee, had exploited a misconfigured web application firewall to access Capital One's data stored on AWS servers. The breach exposed personal information from credit card applications submitted between 2005 and early 2019, including names, addresses, phone numbers, email addresses, dates of birth, and self-reported income. About 140,000 Social Security numbers and 80,000 linked bank account numbers were also compromised.

In response to the incident, Capital One faced several legal and regulatory actions:

Regulatory Fines: In August 2020, the Office of the Comptroller of the Currency (OCC) fined Capital One $80 million for failing to establish effective risk assessment processes before migrating operations to the cloud.

Class-Action Settlement: In 2022, Capital One agreed to a $190 million settlement to resolve a class-action lawsuit filed by affected customers. The settlement provided compensation for out-of-pocket losses and offered identity protection services.

Among the many lessons learned from the Capital One Breach is the importance of ensuring the entirety of the IAM estate (human, non-human, FTEs, Consultants, perms, as well as temps) is fully understood, access controls have limited exposure and, of course, the importance of managing both human and machine identities, particularly in cloud environments.

The Expanding Scope of Identity in Cybersecurity

In today's digital ecosystem, Identity and Access Management (IAM) is a foundational security function encompassing more than just user accounts. Modern IAM now addresses two distinct categories of identities: human and non-human. As organizations adopt digital transformation, integrate cloud-based systems, and expand the use of automation and IoT, the identity management landscape has grown increasingly complex. For CISOs, knowing their IAM estate across both human and non-human identities is critical for maintaining security, regulatory compliance, and operational efficiency.

This chapter delves into why CISOs must understand the IAM estate comprehensively, how the rise of non-human identities impacts cybersecurity strategies, and how an integrated approach can help CISOs address emerging risks and challenges.

The Evolving IAM Landscape: From Human Users to Non-Human Identities

IAM traditionally focused on managing human identities, mainly employees accessing internal resources. However, the shift toward cloud environments, remote work, third-party collaboration, and automated processes has introduced a new set of non-human identities. These include service accounts, application identities, APIs, and IoT devices. The significance of these non-human entities has grown substantially, as they often operate autonomously and require high levels of access to critical systems.

Today's IAM estate, therefore, requires:

- **Comprehensive Management of Both Identity Types:** Ensuring security, efficiency, and regulatory compliance for human and non-human users.
- **Advanced Access Controls:** Implementing granular access management, especially as non-human identities frequently possess elevated permissions that could lead to catastrophic breaches if misused.
- **Continuous Monitoring and Auditing:** As both identity types interact with sensitive systems, monitoring activity is vital to identify potential anomalies and prevent malicious use of credentials.

Understanding this multi-dimensional IAM estate enables CISOs to take a holistic approach to security, addressing risks associated with both human and non-human identities.

Human Identities: Managing People-Centric Access Needs

Human identities represent traditional user accounts associated with employees, contractors, partners, and customers. These identities are subject to dynamic access requirements and pose unique security challenges. Critical components of human identity management are:

- **Role-Based Access Control (RBAC):** Assigning access based on roles allows CISOs to enforce the principle of least privilege and ensure that users can only access resources relevant to their job functions.
- **Authentication Methods:** Multi-Factor Authentication (MFA) and adaptive authentication enhance security by requiring additional verification steps, especially for sensitive systems.
- **Lifecycle Management:** Effective lifecycle management ensures access rights are updated in real-time as users change roles, join or leave the organization, or change their job responsibilities.

Risks Associated with Human Identities

Human identities carry significant risk due to potential insider threats, phishing attacks, and social engineering tactics that target individual users to gain unauthorized access. Key challenges include:

- **Credential Compromise:** Stolen or compromised passwords remain a top risk factor for data breaches.
- **Over-Privileged Accounts:** Users often retain access rights beyond their actual needs or tenure, increasing the likelihood of misuse.

- **Unmanaged Third-Party Access:** Contractors and partners may require access to internal systems, but managing their access securely is often overlooked, leading to potential security gaps. By understanding and closely managing human identities, CISOs can implement tailored security measures, such as least-privilege access, regular access reviews, and robust authentication protocols that reduce human-related risks.

Non-Human Identities: Managing Machine-Driven Access in Modern Enterprises

Non-human identities, also known as machine identities, are created to enable automated processes, application interactions, and IoT/OT device operations. These identities often have privileged access levels and perform essential functions, but they also present unique security challenges. Types of non-human identities are:

- **Service Accounts:** These are used by applications or services to connect and exchange data. They typically have elevated permissions and are less frequently monitored.
- **Application Identities:** Many applications require their own identity to interact with resources, often in cloud environments, and require access to sensitive data and APIs.
- **Robotic Process Automation (RPA) Bots:** RPAs automate tasks by mimicking human actions and require access to systems and databases, often needing elevated permissions.
- **IoT/OT Devices:** The Internet of Things and Operational Technology devices, such as sensors and industrial equipment, increasingly interact with enterprise networks, requiring distinct identity management and security controls.

Risks Associated with Non-Human Identities

The rapid proliferation of non-human identities poses a distinct set of risks:

- **Privilege Overload:** Non-human identities often have access to critical systems and data but lack appropriate privilege management controls.
- **Credential Mismanagement:** Non-human accounts frequently have static or weak credentials, which, if exposed, could provide attackers with a direct route into critical systems.
- **Lack of Visibility:** Non-human identities operate autonomously, often leading to "identity sprawl," where unmonitored or unaccounted identities accumulate within systems. Failure to manage non-human identities effectively can lead to significant vulnerabilities, as attackers increasingly target machine identities to exploit system-level access. For CISOs, achieving visibility and control over these identities is crucial for minimizing the potential attack surface.

Why CISOs Must Know Their Entire IAM Estate

A well-understood IAM estate allows CISOs to apply robust security controls across all identity types, creating a secure, compliant, and efficient access framework. Key benefits include an enhanced security posture. Understanding the IAM estate enables CISOs to enforce Zero Trust principles, which assume no user or device is trusted by default, regardless of location within or outside the network.

With a Zero Trust approach the following can be implemented:

- **Continuous Authentication and Verification:** Regularly authenticating both human and non-human identities.

- **Least Privilege Enforcement:** Ensuring every identity has only minimal access required, reducing exposure to sensitive systems and data.
- **Privileged Access Management (PAM):** Implementing PAM solutions helps protect, monitor, and control access to critical systems by privileged accounts—human or machine. Operational Efficiency and Reduced Identity Sprawl Aa well-managed IAM estate and avoids identity sprawl, where unused, untracked, or redundant identities accumulate. CISOs can:
- **Automate Identity Provisioning and De-Provisioning:** Streamline the onboarding and offboarding process for human and non-human identities.
- **Regular Access Reviews:** Periodic reviews of access rights help eliminate excess permissions, especially for non-human identities.
- **Centralized Management:** Centralized IAM platforms provide a single pane of glass for visibility, allowing for consistent and efficient identity management across all systems. Regulatory Compliance and Risk Management Regulatory frameworks (e.g., GDPR, HIPAA, SOX) mandate strong IAM practices, including detailed access controls, audit logs, and regular reviews.

By understanding the IAM estate, CISOs can ensure:

- **Regulatory Alignment:** Meeting compliance requirements through proper access governance and documentation.
- **Incident Response Preparedness:** Knowing which identities access which systems helps CISOs respond effectively to incidents, enabling swift remediation of potential breaches.
- **Board and Stakeholder Reporting:** CISOs can communicate security posture and compliance efforts more effec-

tively to the board and other stakeholders by maintaining comprehensive IAM oversight.

Critical Practices for CISOs Managing Human and Non-Human Identities:

- **Implement Strong Authentication for All Identities:** MFA and adaptive authentication for human and non-human identities significantly strengthen security.
- **Use Privileged Access Management (PAM) solutions:** PAM solutions are critical for controlling and monitoring elevated access, ensuring the security of both human and non-human privileged accounts.
- **Monitor and Audit Continuously:** Advanced analytics and UEBA solutions help detect abnormal behaviors in both identity types, allowing for proactive threat detection.
- **Enforce Lifecycle Management:** Automating the provisioning, de-provisioning, and ongoing management of identities ensures up to date access rights.
- **Apply Segmentation and Micro-Segmentation:** Limiting network access based on identity type, usage patterns, and necessity helps reduce risk and enforces Zero Trust principles.

Conclusion

The Strategic Value of a Comprehensive IAM Estate. Identity forms the backbone of security, so CISOs must prioritize understanding their IAM estate across humans and machines. As identities increasingly become the primary access point in digital ecosystems, this insight provides a robust defense against threats. It also improves operational efficiency and ensures regulatory compliance. Knowing the IAM estate inside and out is no longer an option for CISOs; it is a strategic necessity. By embracing a comprehensive, identity-focused approach, CISOs can secure the organization, empower growth, and maintain stakeholder trust in the ever-evolving digital landscape.

Chapter 7

KNOW THY CYBER RISK POS-
TURE

Importance of Knowing The Attack Surface - Know Thy Cyber Risk Posture

K nowing what is being protected is critical for CISOs, and understanding the posture of the assets being protected is equally important.

To that end, a robust Security Vulnerability Posture Management (SVPM) is vital. Continuously assessing, prioritizing, and proactively mitigating security vulnerabilities within an organization's IT environment is foundational to the success of any cyber program. It involves identifying weaknesses in Software, systems, networks, and applications and taking proactive measures to address these vulnerabilities.

For CISOs, the importance of SVPM can be outlined in several key areas:

- **Risk Mitigation:** SVPM allows organizations to proactively identify and mitigate security vulnerabilities before malicious actors can exploit them. By addressing vulnerabilities promptly, cyber defenders can significantly reduce the risk of cyberattacks, data breaches, and unauthorized access. This proactive approach is essential for maintaining a robust security posture.

- **Compliance and Regulations**: Many industries and regions have specific regulations and compliance standards related to cybersecurity. SVPM helps organizations comply with these regulations by identifying vulnerabilities and ensuring necessary security measures are in place. Compliance with industry standards helps avoid legal consequences and demonstrates a commitment to data security and customer privacy.
- **Prioritization of Remediation Efforts:** Not all vulnerabilities are created equal; some pose a higher risk to an organization's security than others. SVPM tools analyze vulnerabilities and prioritize them based on severity, potential impact, and exploitability. This prioritization helps security teams focus their remediation efforts on addressing the most critical vulnerabilities first, ensuring that resources are allocated effectively.
- **Patch Management**: SVPM plays a crucial role in patch management processes. It helps organizations identify which systems and applications are vulnerable and require patching. By automating the identification of vulnerable assets, cyber defenders can streamline patch deployment processes, reducing the window of exposure to known vulnerabilities and enhancing overall security.
- **Incident Response and Forensics**: A clear understanding of the organization's vulnerability posture is essential in a security incident. SVPM provides valuable data for incident response and forensics efforts. Security teams can quickly assess which vulnerabilities might have been exploited, understand the attack vector, and take appropriate measures to

contain the incident and prevent similar attacks in the future.

- **Third-Party Risk Management:** Organizations often rely on third-party vendors and partners whose systems and applications are interconnected. SVPM helps assess the security posture of these external entities by identifying vulnerabilities in their products or services. This information is crucial for evaluating and managing third-party risks, ensuring that partners adhere to security best practices.

- **Continuous Improvement**: Cyber threats and vulnerabilities are constantly evolving. SVPM is not a one-time activity but a constant process that adapts to new threats and vulnerabilities as they emerge. By continuously monitoring the security vulnerability posture, organizations can stay ahead of cyber threats, adjust their security strategies, and implement necessary countermeasures to protect their digital assets effectively.

Robust Security Vulnerability Posture Management is therefore indispensable to CISOs for mitigating risks, ensuring compliance, prioritizing remediation efforts, efficient patch management, aiding incident response, managing third-party risks, and enabling continuous improvement in an organization's cybersecurity defenses.

Whereas SVPM is typically focused on inside-the-perimeter digital assets, the 'outside in' security posture of an organization's assets is critical to protecting its digital estate. This posture is typically referred to as External Attack Surface Management (EASM), which refers to the process of identifying, monitoring, and managing an organization's digital footprint and online presence, which is accessible from outside the organization's network. This includes all publicly accessible assets, such as websites, domains, sub-domains, APIs, and cloud services.

Chapter 8

KNOW THY EXTERNAL ATTACK SURFACE

Importance of Knowing Thy External Attack Surface

The 2017 Equifax data breach, cited as one of the most significant cybersecurity incidents to date as it exposed sensitive personal information of approximately 147.9 million individuals, which held several important lessons for me. The breach resulted from exploiting a known vulnerability in the Apache Struts web application framework, specifically CVE-2017-5638. This flaw allowed remote code execution on affected systems. The Apache Software Foundation released a patch for this vulnerability on March 7, 2017. Despite this, Equifax did not apply the patch promptly, leaving their systems exposed until May of 2017, when their systems were compromised, though it took until September 2017 for Equifax to announce the breach publicly. After exploiting the Apache Struts vulnerability, attackers executed remote commands to infiltrate Equifax's network. They moved laterally within the network, escalating privileges and accessing databases containing personal information. The attackers maintained access for 76 days before detection. In February 2020, the U.S. Department of Justice indicted four China's People's Liberation Army members, accusing them of orchestrating the Equifax breach. Among the many lessons I took away from this breach are the following:

- **Ramifications of delayed Patch Management:** Equifax failed to promptly apply the Apache Struts patch, leaving systems vulnerable for months.
- **Ineffective Vulnerability Scanning:** The scans did not identify the unpatched systems, indicating deficiencies in the scanning process.
- **Insufficient Network Segmentation:** Attackers could move laterally within the network, suggesting inadequate segmentation and access controls.

There's something to be said for "CISO's Circle of Trust" since the primary and secondary vulnerability scanners deployed by my team at the Federal Reserve at the time also failed to detect the Struts vulnerability on one of the Fed's Internet-facing sites until I received a call from a CISO colleague at one of the regulated banks highlighting this 'miss' by our scanners.

It goes without saying, then, that managing the external attack surface is critically important for several reasons:

- **Vulnerability Identification:** By comprehensively mapping the external attack surface, cyber defenders can identify potential vulnerabilities that malicious actors might exploit. Vulnerabilities in publicly accessible systems can be detected and remediated proactively, reducing the risk of cyberattacks, data breaches, and unauthorized access.
- **Risk Assessment and Prioritization:** Understanding the external attack surface allows CISOs to assess and prioritize security risks. Not all assets pose the same level of risk, and by analyzing the attack surface, cyber defenders can focus their efforts on securing the most critical assets first. This risk-based approach ensures that resources are allocated effectively to protect high-value targets.

- **Third-Party Risk Management:** Many organizations rely on third-party vendors and services, each adding to the external attack surface. Managing this surface helps CISOs assess the security posture of their vendors and partners. By monitoring external assets related to third parties, organizations can ensure that these entities adhere to security best practices, reducing the risk of supply chain attacks.
- **Incident Response and Forensics:** A detailed understanding of the external attack surface enables rapid response in a security incident. Cyber defenders can quickly identify compromised assets, understand the scope of the breach, and take necessary actions to contain the incident. This rapid response capability is essential for minimizing the impact of security breaches and accelerating recovery efforts.
- **Compliance and Regulatory Requirements:** Many regulations and industry standards require organizations to maintain a secure online presence. By managing the external attack surface, organizations can demonstrate compliance with regulatory requirements. This is crucial for industries such as finance, healthcare, and government, where strict compliance standards are mandated by law.
- **Brand Protection:** Public-facing assets like websites and social media profiles are often the first contact points between a business and its customers. A compromised online presence can damage a company's reputation and erode customer trust. Managing the external attack surface helps protect the brand by ensuring that online assets are secure, reducing the risk of defacement, phishing attacks, or other malicious activities that could harm the organization's reputation.
- **Threat Intelligence Integration:** EASM allows cyber defenders to integrate intelligence feeds to identify emerging threats and vulnerabilities proactively. Organizations can stay ahead of potential attacks by continuously monitoring

the external attack surface and correlating this information with threat intelligence data, enabling a proactive defense strategy against evolving cyber threats.

External Attack Surface Management should, therefore, be considered a core component of a CISO's cyber strategy based on its criticality in helping identify vulnerabilities, assess risks, manage third-party relationships, facilitate incident response, ensure regulatory compliance, protect brand reputation, and integrate threat intelligence.

A robust EASM's key components should include a robust Bug Bounty and Continuous Penetration Testing (as a Service or Platform). Bug Bounty and PTaaS (Penetration Testing as a Service) platforms enable organizations to continuously test their digital assets for vulnerabilities by leveraging external security researchers, ethical hackers, and automated testing solutions. Here's why these platforms are vital for EASM:

- **Continuous Threat Discovery and Remediation:** Unlike traditional penetration testing, which may occur annually or biannually, Bug Bounty and PTaaS platforms provide continuous testing. They constantly monitor and probe an organization's attack surface, uncovering vulnerabilities as they emerge.
- **Real-World Attack Simulation:** These platforms replicate real-world attack scenarios more closely than automated tools. Security researchers and ethical hackers simulate sophisticated attacks, identifying potential risks that could otherwise remain undetected, especially in large or complex environments.
- **Scalability and Adaptability:** Bug Bounty platforms' crowdsourced nature means organizations can scale vulnerability discovery efforts without needing an in-house team. PTaaS platforms, on the other hand, combine automation with on-demand access to expert testers, providing flexibility that adapts to the organization's evolving attack surface.

- **Cost-Effective Vulnerability Management**: Bug Bounty programs allow organizations to pay only for validated vulnerabilities, making them a cost-effective model for vulnerability management. PTaaS platforms offer subscription-based services that often include continuous monitoring, enabling predictable budgeting for security testing efforts.

Critical Benefits of Bug Bounty and PTaaS in EASM

Integrating Bug Bounty and PTaaS platforms into an EASM strategy provides organizations with several benefits:

- **Broader Visibility Across the Attack Surface:** Bug Bounty programs bring a diverse pool of security researchers with varied expertise and perspectives. This diversity helps identify vulnerabilities across various assets, from web applications and APIs to cloud configurations and IoT devices. PTaaS platforms often include asset discovery and scanning capabilities that map the organization's digital footprint, helping identify unknown or forgotten assets.
- **Faster Vulnerability Detection and Patch Cycles:** Both Bug Bounty and PTaaS platforms support rapid vulnerability detection and disclosure, allowing organizations to patch vulnerabilities sooner. Many PTaaS platforms also offer real-time reporting and collaboration tools that streamline the remediation process, reducing the window of exposure to potential exploits.
- **Enhanced Security for Compliance and Risk Management:** Regular testing through Bug Bounty and PTaaS helps meet regulatory and compliance requirements for continuous monitoring and vulnerability management. By actively managing their attack surface, organizations can

demonstrate a proactive security posture to stakeholders, regulators, and customers, ultimately reducing risk.

- **Community and Industry Intelligence:** Bug Bounty platforms attract a global community of security researchers, exposing organizations to industry-wide intelligence and emerging threat patterns. PTaaS platforms often include threat intelligence feeds and report trending vulnerabilities, giving organizations additional insights into the broader threat landscape.

Implementing Bug Bounty and PTaaS for Effective EASM

To successfully integrate Bug Bounty and PTaaS platforms into an EASM strategy, CISOs should consider the following steps:

- **Identify Key Assets and Prioritize Testing:** First, map critical assets within the attack surface that need regular testing, such as public-facing applications, APIs, and sensitive databases.
- **Define Program Scope and Rules of Engagement:** Specify which assets are in scope, set reporting guidelines, and outline acceptable testing methods. Clear rules of engagement ensure that Bug Bounty and PTaaS participants understand organizational expectations and security requirements.
- **Set Reward Structures for Bug Bounty Programs:** Determine reward tiers based on the severity of vulnerabilities found. A well-structured incentive model attracts skilled researchers and encourages thorough testing.
- **Establish a Responsive Remediation Process:** Develop a process for quickly assessing and patching reported vulnerabilities. Bug Bounty and PTaaS platforms typically in-

clude workflow tools for triaging reports, allowing security teams to prioritize and address issues in real-time.

- **Integrate Findings with EASM Tools:** Many organizations use EASM tools to monitor their attack surface. By integrating Bug Bounty and PTaaS findings with these tools, CISOs can create a centralized view of vulnerabilities, enabling a more cohesive and streamlined EASM strategy.

Chapter 9

KNOW THY CMDB

Know Thy CMDB - The Need for Configuration Management

C onsolidating all these component inventories (Hardware, Software, Data, SBOM, API, etc.) should be done using an authoritative Configuration Management Database (CMDB). This centralized repository stores information about an organization's IT assets and configurations, typically providing a detailed record of hardware, Software, network components, and their relationships within the IT environment. It's impossible for CISOs to adequately protect their organization's assets without knowing the assets within the organization they are charged with protecting. The critical importance of a CMDB for cyber defenders and IT operators is highlighted in several key areas as follows:

- **Visibility and Control:** A CMDB offers comprehensive visibility into the IT infrastructure. It provides cyber defenders with an up-to-date and accurate view of all assets, configurations, and interdependencies. This visibility is crucial for understanding the IT landscape, enabling informed decision-making, especially during cyber exposure analysis or business impact analysis determinations following a cyber breach.

- **Change Management:** Effective change management is a cornerstone of IT service management. A CMDB helps organizations track changes made to their IT assets over time. By recording changes and their impact on configurations, IT teams can assess potential risks, plan for contingencies, and ensure that changes do not disrupt critical services. It promotes a systematic and organized approach to change management processes.

- **Incident and Problem Management**: Accurate information about the configuration items involved is essential during incident resolution and problem analysis. A CMDB helps IT support teams quickly identify the affected components, their configurations, and their relationships with other elements in the infrastructure. This accelerates the resolution process, reduces downtime, and enhances user satisfaction by minimizing service disruptions.

- **Asset Lifecycle Management:** IT assets have a lifecycle that includes procurement, deployment, maintenance, and retirement phases. A CMDB allows organizations to track the entire lifecycle of assets, from acquisition to disposal. This comprehensive view helps optimize asset usage, plan upgrades, and ensure compliance with licensing agreements. Efficient asset lifecycle management leads to cost savings and better resource utilization.

- **Capacity and Performance Management:** Understanding the configuration and capacity of IT resources is crucial for managing performance and ensuring optimal utilization. A CMDB provides detailed information about hardware specifications, software versions, and network configurations. IT teams can use this data to analyze performance trends, plan for capacity upgrades, and optimize resource allocation, leading to improved system performance and user experience.

- **Compliance and Audit Readiness**: Many industries have regulatory requirements for IT asset management and data security. A CMDB helps organizations demonstrate compliance by providing a comprehensive record of configurations and changes. A well-maintained CMDB simplifies validating compliance with industry standards and regulations during audits.
- **Disaster Recovery and Business Continuity**: In a disaster, organizations must quickly assess the impact on their IT infrastructure and plan for recovery. A CMDB is a critical reference point for identifying affected systems and applications. It supports disaster recovery planning by ensuring that organizations can prioritize restoring essential services and minimizing downtime.

A Configuration Management Database (CMDB) is equally critical for CISOs and IT operators because it provides visibility, control, and accurate information about IT assets and configurations. It supports efficient change management, IT and Cyber incident resolution, problem analysis, asset lifecycle management, capacity planning, compliance efforts, and disaster recovery initiatives.

KNOW THY THIRD PARTIES

Know Thy Third Parties – The Strategic Significance Of Third Party Risk Management

In my journey across industries as a CISO—whether in government, finance, or elsewhere in the private sector—one thing has remained consistent: the critical need for third-party solid risk management. As organizations grow more interconnected and reliant on external vendors, third-party relationships become enablers of innovation and gateways to potential disaster. We've seen time and again how vulnerabilities in third-party systems can unravel the security fabric of even the most fortified enterprises. The Panama Papers, Snowden revelations, SolarWinds, MOVEit, HashiCorp, Kaseya, and the HPE/DXC cloud breach are stark reminders that third-party risk management is important for CISOs today.

The Expanding Digital Ecosystem

Businesses today are part of vast digital ecosystems, sharing data, systems, and infrastructure with external providers. The reliance on third parties has grown exponentially from cloud services to managed security providers to niche software vendors. In my career, I've seen how this shift has brought businesses significant efficiencies, scalability, and agility. However, it also brings an expanded attack surface—a surface where, often, as your organization's principal cyber defender, you do not have direct visibility or control.

In healthcare, for instance, third parties manage patient data and ensure critical systems are always operational. In finance, they handle sensitive transactions and personal financial data. Third parties often manage sensitive classified information in government and even the military, where I've spent a significant part of my career. Across all these industries, the need to trust external vendors and contractors is paramount. Still, trust must be earned, verified, and continuously monitored consistently with the "trust but verify" foundational principles.

Panama Papers: Data in the Wrong Hands

The Panama Papers leak of 2016 offers a textbook example of how a breach by a third party can have devastating consequences. Over 11.5 million confidential documents from Mossack Fonseca, a Panamanian law firm, were leaked, exposing the offshore dealings of individuals and corporations worldwide. The law firm was a trusted third party that handled sensitive financial and legal records for powerful entities, but their security controls were inadequate.

This incident reinforced that it's not enough to secure our internal systems. As CISOs, we must scrutinize every vendor that handles sensitive information on our behalf. Even the most seemingly secure partnerships can be the source of catastrophic leaks if not properly vetted. This incident taught me that diligence in assessing third-party data protection protocols is just as important as securing our systems, particularly when dealing with legal or financial entities that are seen as attractive targets.

Snowden Revelations: The Insider Threat Amplified

The Edward Snowden breach, widely regarded as one of the most significant intelligence leaks in U.S. history, involved the unauthorized disclosure of classified National Security Agency (NSA) documents by former NSA contractor Edward Snowden in 2013. Snowden's actions exposed extensive details about U.S. and allied surveillance programs, raising serious questions about insider threat management, third-party risk, and government oversight.

Overview of the Snowden Breach

Context: Edward Snowden was an IT contractor working for Booz Allen Hamilton, which held contracts with the NSA. Due to his role, he had broad access to classified information.

Disclosure: Snowden collected and leaked a massive trove of NSA documents to journalists, revealing surveillance programs like PRISM, which monitored global communications data, and other programs focused on collecting metadata from telecommunications networks.

Impact: The breach led to global political fallout, substantial changes in government surveillance laws, and a renewed focus on privacy rights, as well as lasting implications for security and risk management within government and private sectors.

Implications for CISOs in Two Key Areas

1. **Third-Party Risk Management (TPRM) Program** - Snowden's role as a contractor at Booz Allen Hamilton brought significant attention to the risks of third-party vendors in securing sensitive information. The incident exposed gaps in oversight for third-party employees working on government projects and emphasized the importance of extending cybersecurity governance beyond internal staff.

   Essential Lessons and Best Practices for CISOs in TPRM are listed below:
   - **Enhanced Vetting of Third Parties:** Organizations must implement rigorous background checks, especially for contractors with access to sensitive information. This includes reviewing criminal records, financial history, and prior employment to screen for potential risk factors.
   - **Detailed Access Control Policies:** CISOs should ensure that third-party access is managed on a need-to-know basis. Rather than granting broad, unrestricted access, they should enforce granular controls tailored to the specific needs of each third-party role.
   - **Continuous Monitoring and Auditing**: It's not enough to vet third parties during onboarding. CISOs should establish ongoing monitoring for third-party activities and usage patterns, focusing on vendors handling sensitive information. Automated alerts for unusual access or behavior patterns can help detect risks early.
   - **Third-Party Security Assessments and Audits:** Regular security assessments, such as SOC 2 or ISO 27001 certifications, help verify that vendors meet security standards. CISOs should require periodic audits to ensure third parties maintain and comply with security controls over time.
   - **Data Segmentation and Least Privilege:** Limiting access to only the necessary information is crucial. Organizations

should employ data segmentation and enforce the principle of least privilege so that third parties can only access data pertinent to their roles.

- **Incident Response Coordination:** Establish clear protocols for coordinating incident response activities with third-party vendors. This includes defining roles, responsibilities, and notification timelines in case of a breach. A well-coordinated response helps mitigate the impact of third-party incidents.

The Snowden breach demonstrated the need for third-party solid risk management practices, especially when third-party personnel have access to sensitive data or critical systems. By proactively managing these risks, CISOs can reduce the potential for third-party breaches and maintain tighter control over sensitive information.

2. **Insider Risk Management (IRM) Program** - The Snowden incident was a quintessential insider threat case, where a trusted individual exploited access to sensitive data for unauthorized purposes. This has made insider risk management a top priority for CISOs, particularly in organizations with high-stakes data or government contracts.

Essential Lessons and Best Practices for CISOs in IRM are:

- **Comprehensive Access Controls and Monitoring:** Organizations must monitor all employees and contractors for unusual activity. This includes implementing robust logging and monitoring tools that track access to sensitive information and detect anomalies, such as accessing large volumes of data or systems outside of regular hours.

- **Behavioral Analytics:** Modern insider threat programs use behavioral analytics to detect potentially risky behavior patterns. For instance, sudden spikes in data access, irregular login locations, or bypassing usual access routes can trigger alerts and prompt further investigation.

- **Data Loss Prevention (DLP) Tools:** DLP solutions monitor and restrict data transfer activities, such as copying files to external drives or uploading to cloud storage. These tools can help enforce policies that prevent employees from exfiltrating sensitive information.

- **Zero-Trust Security Model:** A zero-trust approach, where no user is inherently trusted and access is continually verified, can help mitigate insider risk. This approach requires strong authentication, authorization, and continuous monitoring to detect real-time unusual activity.

- **Periodic Access Reviews:** Regularly reviewing and re-assessing employees' access to sensitive information helps reduce unnecessary privileges. This includes adjusting access based on role changes or employment status updates, particularly for individuals transitioning from sensitive roles.

- **Clear Policies and Training on Data Security:** Policies should be communicated to all staff, including contractors, about the consequences of unauthorized data access. Training programs should include information on reporting suspicious behavior and reinforce the importance of data security.

- **Encourage Whistleblowing and Support Reporting Mechanisms:** Establishing an anonymous reporting system allows employees to report concerns about suspicious behavior without fear of retaliation. CISOs should encourage a culture where employees feel comfortable reporting potential insider threats.

The Snowden incident highlighted the risks posed by insiders with extensive access to sensitive data and the importance of proactive detection and response mechanisms. By developing a comprehensive IRM program, CISOs can reduce the likelihood of insider breaches and better safeguard organizational data. The Snowden breach underscores critical aspects of risk management for CISOs, particularly in managing third-party access and mitigating insider threats. To protect against similar incidents, CISOs should develop comprehensive TPRM and IRM programs that enforce strong access controls, utilize monitoring tools, and foster a culture of security across all employees and contractors.

SolarWinds and Kaseya: Supply Chain Vulnerabilities Exposed

The SolarWinds breach in 2020 and the Kaseya ransomware attack in 2021 were among the most damaging supply chain attacks in recent history. Both incidents involved malicious actors exploiting vulnerabilities in widely used software platforms to gain access to thousands of downstream customers. In the case of SolarWinds, Russian state actors injected malware into the company's Orion software updates, compromising major government agencies and private organizations. On the other hand, Kaseya saw ransomware operators exploiting a vulnerability in its Virtual Systems Administrator (VSA) management software, leading to widespread ransomware infections.

For me, these incidents underscored how crucial it is to assess and mitigate the risks associated with software supply chains. When third-party software is deeply integrated into your network, it's not just an external risk—it becomes an internal one. These incidents reinforced my approach to continuously assess vendor software for vulnerabilities and conduct thorough security audits, layered security controls, network segmentation, and redundant security controls to mitigate the risk of a single point of failure.

MOVEit: The Importance of Data Transfer Security

The MOVEit breach, which became widely known in mid-2023, was a significant cyber incident involving the exploitation of vulnerabilities in the MOVEit Transfer software, a file transfer and managed file transfer (MFT) solution developed by Progress Software. Organizations widely use this software for secure file sharing and data transfers.

The breach exploited a zero-day vulnerability (CVE-2023-34362) in the MOVEit Transfer software, allowing attackers to gain unauthorized access to databases, download sensitive files, and potentially install web shells to maintain persistent access. This vulnerability was explicitly in the MOVEit Transfer product and allowed SQL injection, enabling attackers to bypass authentication and escalate privileges. The breach was first discovered in late May 2023 and was being actively exploited before it was publicly disclosed, indicating a period of undetected exposure. Shortly after the disclosure, Progress Software released patches to address the vulnerabilities, but many organizations had already been impacted by that point. The ransomware group CL0P (or Clop), a cybercriminal gang linked to other notable ransomware attacks, claimed responsibility for the MOVEit attacks. CL0P is known for exploiting zero-day vulnerabilities in widely used software products and extorting companies by threatening to release stolen data. The attackers used SQL injection to access MOVEit's databases, retrieving sensitive information stored by client organizations. Once they accessed the systems, they exfiltrated large volumes of data and installed web shells, allowing them to re-enter systems even after the initial vulnerability was patched.

The MOVEit breach impacted hundreds of organizations globally, spanning multiple industries, including finance, government, education, healthcare, and retail. Sensitive information, including Personally Identifiable Information (PII), financial records, healthcare data, and internal corporate documents, was compromised for many affected organizations. This poses a significant privacy and regulatory risk, especially for entities in highly regulated industries like finance and healthcare. The incident also increased regulatory scrutiny on software providers and third-party risk management practices, especially for solutions handling sensitive data.

The MOVEit breach highlighted the risks associated with third-party software providers, especially for critical services like file transfers. Organizations are increasingly evaluating their third-party risk management practices to prevent similar incidents in the future. Given the international impact, the breach brought regulatory bodies' attention to third-party vulnerabilities and the security of widely used applications. This incident may prompt more stringent requirements for vendors and software providers to address vulnerabilities promptly and transparently.

The MOVEit breach remains a crucial example of the growing risks associated with third-party software vulnerabilities, underscoring the importance of robust cybersecurity practices across industries.

Building a Resilient Third-Party Risk Management Strategy

From these high-profile incidents, I've understood that the relationship with third-party vendors must be more than transactional —it must be strategic. A robust third-party risk management program involves several key elements that I've implemented across different organizations over the years:

1. **Rigorous Due Diligence:** Vendors must undergo thorough due diligence before any engagement. This includes assessing their security controls and evaluating their business practices, financial health, and past performance. Getting a complete picture of who you're partnering with is essential.

2. **Contractual Controls:** Security requirements must be clearly articulated in contracts. These should include expectations around incident reporting, data protection protocols, audit rights, and breach response. A well-drafted contract can mitigate risks by holding vendors accountable for security failures.

3. **Continuous Monitoring:** Once a vendor is onboard, the work doesn't stop. Regular audits, assessments, and monitoring tools should be in place to keep track of their security posture. This can include anything from automated vulnerability scans to annual third-party security assessments.

4. **Incident Response Integration:** Every third party must be integrated into the organization's incident response plan. This means having clear lines of communication and predefined procedures for when things go wrong. Vendors must know their role in your incident response strategy and be prepared to act swiftly in case of a breach.

5. **Limiting Privileged Access:** As the Snowden incident taught us, privileged access must be limited. Vendors should only have access to the systems and data necessary to perform their duties, and this access should be continuously reviewed and revoked when no longer needed.

Third-Party Risk is Everyone's Problem

Across all industries, from government to finance, energy, and healthcare, third-party risk remains one of the most significant and challenging aspects of cybersecurity management. The high-profile incidents we've discussed are a stark reminder that even the most secure organizations can be threatened by vulnerabilities in their third-party relationships.

As CISOs, we cannot afford to overlook these risks. It's our responsibility to ensure that third-party vendors are held to the same high standards as our internal teams and that we have the tools and processes to manage these relationships effectively. A successful third-party risk management program can mean the difference between a minor security hiccup and a significant breach that threatens the entire organization.

In a world where the digital ecosystem continues to expand, mastering third-party risk management is necessary and a cornerstone of the modern CISO's mandate.

Chapter 11

KNOW THY GEOPOLITICAL LANDSCAPE

Beyond Bits and Bytes: The Strategic Value of Geopolitical Awareness For CISOs

T he Bangladesh Central Bank heist in 2016, where hackers attempted to steal nearly $1 billion from the Central Bank's account at the Federal Reserve Bank of New York, wasn't just about financial gain but a direct consequence of the geopolitical landscape. North Korea, heavily sanctioned by the international community, had limited revenue sources. The regime turned to cybercrime to fund its operations, circumventing the financial restrictions imposed by sanctions. This incident underscored the intersection between geopolitics and cybersecurity, illustrating how nation-states under economic duress might leverage cyber capabilities to achieve their strategic objectives.

The incident was also a refresher of what happened between 2011 and 2013 in the wake of sanctions imposed on Iran when that country launched cyberattacks targeting entities in the United States and its allies. Responding to economic and political pressures, Iranian cyber teams launched sophisticated campaigns against financial institutions, energy companies, and other critical infrastructure. These attacks were not random but calculated moves designed to retaliate against the sanctions and demonstrate Iran's cyber capabilities.

More specifically, between 2011 and 2013, Iranian state-sponsored hackers conducted a series of Distributed Denial of Service (DDoS) attacks against major US financial institutions. These attacks were perceived as retaliation for Western economic sanctions and cyber operations targeting Iran's nuclear program. The campaign, known as Operation Ababil, was orchestrated by the Izz ad-Din al-Qassam Cyber Fighters, a group linked to the Iranian government. The attacks targeted the public-facing websites of nearly 50 US financial institutions, including Bank of America, JPMorgan Chase, and Wells Fargo. By overwhelming these websites with excessive traffic, the attackers disrupted online banking services, preventing customers from accessing their accounts and costing the banks millions in remediation efforts. US officials believe these DDoS attacks were in retaliation for economic sanctions imposed on Iran due to its nuclear activities. Additionally, they were seen as a response to cyber operations like the Stuxnet virus, which targeted Iran's nuclear facilities. In 2016, the US Department of Justice indicted seven Iranian nationals associated with the Islamic Revolutionary Guard Corps for their roles in these attacks.

For CISOs, these events are a stark reminder that the cyber threat landscape is deeply intertwined with global geopolitical dynamics. The rise of state-sponsored cyber activities in response to international sanctions and conflicts has created a complex environment where traditional cybersecurity measures alone are insufficient. Understanding geopolitics is essential for developing robust cyber defense strategies that anticipate and mitigate these nation-state threats.

In the rapidly evolving digital landscape, Chief Information Security Officers (CISOs) shoulder the responsibility of safeguarding organizations from many cyber threats. While their technical expertise remains pivotal, a deep understanding of the geopolitical landscape has emerged as a critical asset. This chapter delves into why CISOs should possess a nuanced comprehension of global geopolitics and how this knowledge enhances their role as cybersecurity guardians.

Geopolitical Influence on Cyber Threats: Anticipating State-Sponsored Attacks

Geopolitical events have a profound impact on the cyber threat landscape. A CISO who understands geopolitical dynamics can anticipate threats from state-sponsored actors, hacktivist groups, and international cyber conflicts. This proactive approach allows organizations to fortify defenses against emerging threats.

The recent Russia-Ukraine conflict serves as a stark reminder of how geopolitical tensions can escalate into widespread cyber warfare. In the lead-up to and during the conflict, Ukraine has been the target of numerous cyberattacks attributed to Russian state-sponsored groups. These attacks have included the deployment of wiper malware designed to destroy data, distributed denial-of-service (DDoS) attacks aimed at crippling critical infrastructure, and sophisticated phishing campaigns targeting Ukrainian government officials and military personnel.

The conflict has impacted Ukraine and had ripple effects across Europe and beyond. Western countries supporting Ukraine, including the United States and NATO members, have experienced increased cyber activities, ranging from disinformation campaigns to direct attacks on critical infrastructure. These cyber operations are often designed to sow discord, disrupt military logistics, or retaliate against sanctions imposed on Russia.

For example, in early 2022, a cyberattack on Viasat, a US-based satellite communications provider, disrupted internet services across Europe, particularly affecting Ukraine's communications capabilities just as the conflict intensified. This attack, attributed to Russian cyber actors, demonstrated the potential for collateral damage in cyber warfare and highlighted the risks to organizations far removed from the physical conflict zone.

Similarly, the increased targeting of energy companies, financial institutions, and supply chain networks in Western countries underscores the broader implications of the Russia-Ukraine conflict. The cyber domain has become a battleground where state actors pursue their geopolitical objectives, and these attacks often extend beyond the immediate region of conflict to affect global markets and businesses.

For CISOs, the Russia-Ukraine conflict exemplifies the need for a deep understanding of geopolitical dynamics to anticipate and mitigate cyber threats. The conflict has shown that cyber operations can be part of a broader strategy to weaken adversaries, influence public opinion, and achieve strategic objectives without direct military engagement.

By staying informed about global geopolitical developments, CISOs can better predict where and when these threats might arise and take preemptive actions to protect their organizations. This includes enhancing threat intelligence capabilities, strengthening incident response plans, and engaging in international cooperation to share information and best practices.

Nation-State Threats: Understanding Motivations and Capabilities

State-sponsored cyberattacks transcend borders and often target critical infrastructure and sensitive data. A CISO who comprehends the geopolitical motivations behind such attacks can devise tailored defense strategies and response plans.

In the case of North Korea and Iran, cyber operations were driven by the need to circumvent economic sanctions and assert geopolitical influence. Understanding these motivations enables CISOs to recognize the potential targets within their organizations and implement specific countermeasures. For instance, financial institutions may need to enhance security around SWIFT systems, while energy companies might focus on protecting Operational Technology (OT) environments from disruption.

The ongoing Russia-Ukraine war further illustrates how state-sponsored cyber activities are intricately linked to broader geopolitical strategies. Russia's cyber operations against Ukraine are part of a more significant effort to undermine the country's ability to function during the conflict, disrupt critical infrastructure, and sow chaos and confusion among the civilian population. These operations have included attacks on government networks, power grids, telecommunications systems, and transportation networks—each carefully targeted to maximize disruption and weaken Ukraine's resilience.

One of the most notable cyberattacks during the conflict was using Industroyer 2 malware, which targeted Ukraine's energy sector to disrupt the country's power supply. This attack mirrored the earlier Industroyer attack in 2016, which caused widespread power outages in Ukraine and demonstrated Russia's capability and intent to leverage cyber tools to achieve strategic objectives.

Another significant example of a state-sponsored cyber activity with global repercussions is the NotPetya malware attack in 2017. Initially launched as a cyberattack against Ukraine, NotPetya quickly spread beyond its target, affecting organizations in over 130 countries. The attack crippled significant multinational corporations, including Merck Pharmaceuticals and Maersk, the global shipping giant. NotPetya was designed to look like ransomware but was, in fact, a wiper malware that irreversibly destroyed data on infected systems. The attack caused billions of dollars in damages and severely disrupted global supply chains, highlighting the potential for state-sponsored cyber operations to inflict widespread collateral damage.

Understanding these attacks allows CISOs in similar industries globally to anticipate potential threats and harden their critical infrastructure against similar tactics. For instance, in the aftermath of NotPetya, many organizations realized the importance of segmenting their networks, ensuring regular backups, and having incident response plans specifically tailored to deal with destructive malware.

Additionally, Russian cyber actors have engaged in widespread disinformation campaigns aimed at undermining trust in Ukrainian institutions and creating divisions among the Ukrainian population and its allies. These campaigns often employ social media platforms to spread misleading narratives, manipulate public opinion, and influence political outcomes. For CISOs, particularly those in the public sector or media organizations, understanding these tactics are crucial for developing defenses against information warfare and ensuring the integrity of information.

The motivations behind Russia's cyber activities are deeply rooted in its geopolitical objectives to weaken Ukraine, challenge Western influence, and assert its dominance in the region. These motivations translate into direct threats for organizations in countries supporting Ukraine or those involved in critical industries. CISOs must, therefore, consider not only the immediate technical aspects of cybersecurity but also the broader geopolitical context that drives these threats.

Supply Chain Risk Management: Assessing Geopolitical Vulnerabilities

Global supply chains are susceptible to geopolitical disruptions that impact an organization's cybersecurity posture. A CISO aware of geopolitical tensions can assess and mitigate risks associated with suppliers and partners in potentially unstable regions.

For instance, companies sourcing components from regions under geopolitical strain may face supply chain disruptions or targeted cyberattacks to compromise product integrity. CISOs must evaluate the geopolitical risks associated with their supply chains and develop strategies to ensure continuity, such as diversifying suppliers or implementing stringent cybersecurity requirements for third-party vendors.

Numerous companies faced significant supply chain disruptions during the Russia-Ukraine conflict, particularly in sectors reliant on raw materials and components from Ukraine and Russia. The conflict led to delays and shortages in essential materials, such as neon gas, critical for semiconductor manufacturing. As a result, global technology and automotive companies experienced production slowdowns, demonstrating the vulnerability of supply chains to geopolitical upheaval.

Another example is the increased risk of targeted cyberattacks aimed at compromising the integrity of products within the supply chain. In 2020, the SolarWinds cyberattack, attributed to Russian state-sponsored actors, highlighted the dangers of supply chain vulnerabilities. The attackers inserted malicious code into SolarWinds' software updates, distributed to thousands of customers, including U.S. government agencies and Fortune 500 companies. This sophisticated attack compromised a trusted software provider, underscoring the need for rigorous supply chain security measures.

Business Continuity and Resilience: Preparing for Geopolitical Crises

Geopolitical crises like conflicts or trade disruptions can ripple through cyberspace, impacting digital operations. A CISO who recognizes these risks can collaborate with business leaders to develop contingency plans that ensure continuity despite geopolitical turmoil.

For example, CISOs can work with other executives to establish alternative communication channels, backup systems, and crisis management teams in anticipation of potential cyberattacks linked to geopolitical tensions. This ensures the organization remains operational even if a cyber incident compromises or disrupts primary systems.

CISOs must evaluate the geopolitical risks of their supply chains and develop strategies to ensure continuity and security. This involves several key actions. Integrating geopolitical awareness into cybersecurity strategy is no longer optional—it is a necessity. As the digital landscape continues to evolve, CISOs' ability to anticipate, understand, and respond to the geopolitical dimensions of cyber threats will be critical to safeguarding their organizations' future.

Chapter 12

KNOW THY THREAT INTEL FEEDS

Know Thy Threat Intel Feeds - The Critical Role of Threat Intelligence for CISOs

I n typical military parlance, getting "left of boom" is the timeline of events before an explosion or incident - a period when the command or unit still has a chance to prepare for and avert a crisis.

With direct applicability and extensibility to cyber, "left of boom" in cyber threat intelligence refers to activities and strategies to prevent or mitigate cyber incidents before they happen. It involves proactive measures to identify, assess, and neutralize threats before they lead to a significant cyber event, or "boom". This approach emphasizes intelligence gathering, analysis, and the development of defensive measures to understand potential adversaries, predict their behavior, and bolster defenses.

In practice, "left of boom" includes threat intelligence gathering, vulnerability assessments, risk analysis, and developing response strategies. By understanding indicators of compromise

(IOCs), tactics, techniques, and procedures (TTPs) of threat actors, organizations can take proactive steps such as patching vulnerabilities, improving security protocols, and training staff to minimize risk. It further implies gathering threat intelligence from various sources, such as internal network logs, security tools, open-source intelligence (OSINT), commercial threat intelligence feeds, and industry-specific sharing communities like Information Sharing and Analysis Centers (ISACs). This forward-looking stance helps CISOs and their organizations move from a reactive to a proactive cyber defense posture, ultimately reducing the impact and likelihood of successful cyberattacks.

As cyber threats evolve in scale, sophistication, and frequency, intelligence-driven defense enables CISOs to make informed decisions, manage resources effectively, and protect critical assets.

This chapter explores the significance of threat intelligence, the challenges involved in implementing it, and the impact of frameworks like STIX (Structured Threat Information eXpression) and TAXII (Trusted Automated eXchange of Indicator Information) on enhancing threat information sharing and analysis.

Understanding Threat Intelligence

- At its core, threat intelligence is gathering, analyzing, and interpreting information about current and emerging threats that could impact an organization's security posture. It includes data on attackers' motivations, tactics, techniques, and procedures (TTPs) and provides actionable insights that allow CISOs and their teams to identify, prepare for, and respond to potential threats. Critical elements of threat intelligence include:
- **Strategic Intelligence:** Provides high-level insights into the motivations and capabilities of threat actors, helping CISOs align security strategy with organizational risk.

- **Operational Intelligence:** Gives actionable insights into specific threats, including how and when they might target the organization.
- **Tactical Intelligence:** Involves understanding specific attacker TTPs, which aids in detecting and responding to threats.
- **Technical Intelligence:** Focuses on technical indicators such as I.P. addresses, domains, and malware hashes, which are essential for immediate defensive measures.

There are a number of reasons why threat intelligence are essential for CISOs such as:

A. Proactive Defense and Risk Mitigation

Threat intelligence allows CISOs to adopt a proactive approach to security. Instead of waiting for threats to materialize, CISOs can anticipate and address risks based on insights gathered from past attacks, industry trends, and real-time intelligence feeds.

- **Early Warning:** By identifying potential threats early, CISOs can prioritize security measures, reduce incident response times, and minimize the likelihood of successful attacks.
- **Informed Decision-Making:** With relevant intelligence, CISOs can make informed decisions about resource allocation, allowing their teams to focus on the most pressing threats.
- **Risk Prioritization:** Threat intelligence enables CISOs to categorize threats based on potential impact, helping them direct resources toward protecting high-risk assets and processes.

B. Enhancing Incident Response and Threat Hunting

For incident response teams, threat intelligence is invaluable in speeding up detection, analysis, and response. It allows teams to respond to incidents with relevant context, shortening investigation times and improving response quality.

- **Improved Detection:** Threat intelligence informs SIEM (Security Information and Event Management) and SOAR (Security Orchestration, Automation, and Response) systems, enabling them to detect anomalous activity linked to known threat indicators.
- **Enhanced Threat Hunting:** By leveraging indicators of compromise (IOCs) and TTPs from threat intelligence feeds, threat-hunting teams can proactively search for hidden threats within the environment.
- **Faster Incident Response:** Threat intelligence provides immediate context on active threats, enabling security teams to respond effectively and limit the impact of an incident.

C. Compliance and Regulatory Support

As regulatory requirements around cybersecurity grow, threat intelligence can help CISOs maintain compliance and meet regulatory expectations, especially those related to data protection and breach notification.

- **Enhanced Reporting:** Threat intelligence enables CISOs to create accurate reports for regulators, boards, and auditors, demonstrating that the organization is actively monitoring and mitigating risks.
- **Meeting Standards:** Frameworks such as NIST and ISO emphasize risk-based approaches to security. Threat intelligence supports these standards by allowing for a proactive, informed response to risks.
- **Breach Notification:** Many regulations, such as GDPR and CCPA, require timely breach notifications. With threat intelligence, organizations can detect breaches early, enabling faster notification and minimizing regulatory penalties.

D. Supporting Business Strategy and Continuity

Threat intelligence goes beyond security and aligns with broader business objectives. CISOs can better protect business continuity and support the organization's strategic goals by understanding threats to critical business processes.

- **Aligning Security with Business Risk:** Threat intelligence helps CISOs contextualize cyber threats in terms of business impact, aligning security efforts with the organization's priorities and risk appetite.
- **Protecting Reputation and Customer Trust:** A robust intelligence-driven security program helps CISOs proactively address risks, reducing the likelihood of data breaches that could harm customer trust and brand reputation.
- **Competitive Advantage:** Organizations that leverage advanced threat intelligence often operate with lower risk and more agility, giving them an edge over competitors in today's digitally dependent market.

The Role of STIX and TAXII in Threat Intelligence

To maximize the value of threat intelligence, CISOs need standardized ways to share and integrate threat information across tools, teams, and organizations. Two frameworks—STIX (Structured Threat Information eXpression) and TAXII (Trusted Automated eXchange of Indicator Information)—provide standardized approaches for structuring and exchanging threat intelligence.

A. **Structured Threat Information eXpression (STIX)**

STIX is a standardized language for representing threat intelligence. It was developed to ensure that security professionals can communicate threat data in a consistent and organized format.

- **Unified Format for Threat Data:** STIX standardizes how threat intelligence is represented, making it easier for different teams and tools to interpret and act on threat information.
- **Detailed Descriptions of Threats:** STIX provides rich detail, including threat actors, their motives, capabilities, and specific Tactic, Techniques, and Procedures (TTPs). This level of detail enables CISOs to implement highly targeted defenses.
- **Interoperability:** By adopting STIX, CISOs can ensure that different threat intelligence sources and cybersecurity tools are interoperable, enabling seamless intelligence integration into their security operations.

B. **Trusted Automated eXchange of Indicator Information (TAXII)**

TAXII protocol facilitates the secure sharing of threat intelligence over HTTPs. It complements STIX by providing a means to transport threat information.

- **Efficient Sharing Mechanism:** TAXII allows organizations to share threat intelligence across organizations securely and in near real-time. This facilitates collaboration within sectors and across industries.
- **Streamlined Intelligence Feeds:** Through TAXII servers, CISOs can receive regular updates from threat intelligence providers, keeping their defenses up to date with the latest threat information.
- **Standardized Protocols for Sharing:** With TAXII, organizations can securely exchange STIX-formatted intelligence, enhancing the value of collaboration while reducing the risk of data leakage.

C. **Practical Application of STIX and TAXII for CISOs**

- **Threat Intelligence Sharing:** Using STIX and TAXII, CISOs can exchange threat intelligence with peers, Information Sharing and Analysis Centers (ISACs), and government agencies, contributing to a collective defense.

- **Automated Threat Feeds:** CISOs can leverage TAXII-compatible threat intelligence feeds, integrating them into SIEM and SOAR platforms for real-time, automated analysis and response.
- **Enhanced Situational Awareness:** STIX and TAXII enable CISOs to visualize relationships between threat actors, TTPs, and indicators, giving them a complete view of the threat landscape.

D. **Implementing Effective Threat Intelligence Programs**

   To build a successful threat intelligence program, CISOs must consider critical aspects of planning, integrating, and operationalizing intelligence. Here are some best practices for CISOs implementing threat intelligence:

- **Define Objectives and Scope**: Set clear objectives for the threat intelligence program based on the organization's risk profile, critical assets, and business needs. Decide on the types of intelligence required—strategic, tactical, operational, or technical.
- **Integrate Intelligence Across Security Operations:** Threat intelligence should be woven into every aspect of security operations. Integrate intelligence into SIEM, SOAR, and Endpoint Detection and Response (EDR) platforms for real-time threat analysis.
- **Collaborate and Share Intelligence:** Collaboration is essential for practical threat intelligence. Participate in industry groups, ISACs, and public-private partnerships to share intelligence and gain insights into emerging threats.
- **Prioritize Threat Intelligence Sources:** Evaluate threat intelligence vendors and sources based on relevance, timeliness, and reliability. Prioritize feeds that

provide actionable intelligence tailored to your organization's industry and risk profile.

- **Regularly Update and Test Intelligence Practices:** The threat landscape changes rapidly, and intelligence must evolve to remain relevant. Periodically test and review the effectiveness of intelligence sources, workflows, and response protocols.

Chapter 13

KNOW THY DEFENSES

Knowing the state of your connected enterprise is equally critical as understanding the efficacy of the cyber defenses deployed to protect your organization.

I n January of 2022, Sounil Yu, the former Chief Scientist at Bank of America, published the Cyber Defense Matrix, which has quickly become a tool every forward-thinking cyber defender should have in their toolbox if they genuinely care about the robustness and coverage of their cyber defenses. The Cyber Defense Matrix is a framework designed to help organizations improve their cybersecurity posture by categorizing various cybersecurity activities into different domains. The matrix provides a structured way to think about cybersecurity defense strategies and align security efforts with organizational goals. The matrix consists of six key domains, each representing a different aspect of cybersecurity:

- **Prevention:** This domain focuses on activities aimed at preventing security incidents. It includes access control, encryption, secure configuration, and patch management. Prevention strategies seek to stop cyber threats from gaining a foothold in the organization's systems and networks.

- **Deterrence:** Deterrence activities aim to discourage potential attackers by implementing security awareness training, legal deterrents (like policies and regulations), and threat intelligence sharing. Organizations can deter potential threats by raising the cost of attacks and increasing adversaries' risks.

- **Detection:** Detection involves activities and technologies that help identify security incidents in real-time or near real-time. This domain includes intrusion detection systems, Security Information and Event Management (SIEM) solutions, anomaly detection, and threat hunting. Detection is crucial for the early identification of cyber threats within the organization's network.

- **Containment:** Containment strategies focus on limiting the impact of security incidents once they are detected. This domain includes network segmentation, incident response planning, and endpoint isolation. Containment aims to prevent attackers' lateral movement within the network and minimize the damage they can cause.

- **Response:** Response activities involve the steps taken to mitigate and recover from a security incident's impact. This domain includes incident response plans, communication protocols, legal actions, and system restoration procedures. A well-defined response plan is essential for minimizing downtime and reducing the overall impact of a security breach.

- **Recovery:** Recovery strategies focus on restoring the organization's systems and processes to normalcy after a security incident. This domain includes backup and disaster recovery plans, system rebuilding, and lessons learnedsessions. Recovery efforts aim to ensure business continuity and prevent similar incidents in the future.

The Cyber Defense Matrix provides a structured framework for cyber defenders to assess their cybersecurity posture, identify potential gaps, and develop a more holistic and proactive approach to cybersecurity.

On the converse, to successfully defend, cyber practitioners should also know who and what they are being called upon to defend against. To aid in that endeavor, in 2013, the Mitre Corporation created the Adversarial Tactics, Techniques, and Common Knowledge or MITRE ATT&CK, which provides a comprehensive knowledge base of adversary tactics, techniques, and procedures (TTPs) based on real-world observations. By aligning their cybersecurity strategies with ATT&CK, cyber defenders can enhance their organization's threat detection and response capabilities.

The framework is organized into matrices representing a different operating system or platform, such as Windows, macOS, Linux, and mobile devices. These matrices are further divided into tactics and techniques. Tactics represent the high-level objectives of an attacker, such as Initial Access, Execution, Persistence, Privilege Escalation, Defense Evasion, Credential Access, Discovery, Lateral Movement, Collection, Exfiltration, and Impact. Conversely, techniques are specific methods adversaries use to achieve these objectives under each tactic. Each technique is described in detail, including a description of the technique, examples of how it has been observed in real-world attacks, potential mitigations, and detection methods.

This information is invaluable for cybersecurity professionals, allowing them to understand the tactics employed by adversaries and develop effective strategies to detect, prevent, and respond to

cyber threats. One of the strengths of the MITRE ATT&CK framework is its constant updates to reflect the evolving threat landscape. As new attack techniques emerge, they are added to the
framework, ensuring that cybersecurity professionals have access
to the most current and relevant information to enhance their security measures. ATT&CK is widely used by cybersecurity teams, including security analysts, incident responders, and threat hunters,
to improve their organization's security posture. It provides a standardized way to communicate about cyber threats, enabling organizations to collaborate better, share knowledge, and strengthen their
defenses against cyber-attacks.

Building on the successes of its ATT&CK framework, Mitre
later released the MITRE D3FEND™ (Detection, Denial, and Disruption Framework Empowering Network Defense), which is a
comprehensive knowledge graph that systematically catalogs cybersecurity countermeasures, offering a structured framework for
understanding and implementing defensive techniques against cyber threats which it developed with funding from the National Security Agency (NSA) and serves as a complementary counterpart
to the MITRE ATT&CK® framework.

Key Features of D3FEND include:
- **Standardized Vocabulary:** D3FEND establishes a common language for defensive cybersecurity techniques, facilitating clear communication among professionals and aiding in selecting and implementing appropriate countermeasures.
- **Structured Knowledge Graph:** The framework organizes
  defensive techniques into a graph structure, detailing their
  relationships to offensive tactics, methods, and procedures.
  This organization aids in understanding how specific defenses can mitigate particular threats.
- **Integration with ATT&CK:** By mapping defensive techniques to corresponding adversary behaviors outlined in

ATT&CK, D3FEND enables organizations to tailor their defenses effectively against known attack vectors.

How CISOs should think about applying D3FEND to protect their organizations include:

- **Security Architecture Design:** Security architects can utilize D3FEND to design robust defense strategies by identifying and implementing countermeasures that address specific adversary techniques.
- **Defense Planning:** Organizations can leverage D3FEND to assess their defensive posture, identify gaps, and plan enhancements to their cybersecurity defenses.
- **Community Collaboration:** D3FEND fosters collaboration across the cybersecurity community by providing a shared framework for discussing and developing defensive strategies.

Chapter 14

KNOW THY IRP (INCIDENT RESPONSE PROTOCOLS)

The Strategic Imperative for CISOs to Master Incident Response Plans

Having led more than a handful of incident response efforts throughout my career, there's much truth to the quote attributed to Mike Tyson that reads "Everyone has a plan until they are punched in the face." The quote is recast in military parlance as "No battle plan ever survives first contact with the enemy" and is attributed to military strategist Helmuth von Moltke. The fitting parallels to cyber incident response, of course, is that while having a structured plan is essential, the reality of a cyber incident is often messier and less predictable than anticipated, but without a well-documented and fully exercised plan, even more chaos will ensue.

Here's how this translates to cyber incident response:

- **Adaptability Under Pressure:** Tyson's quote highlights the importance of being adaptable. During a cyber incident, things rarely go as planned. Attackers change tactics, new vulnerabilities emerge, and unexpected impacts surface. An effective response plan includes flexibility, allowing the team to pivot quickly and respond to unforeseen challenges.

- **Stress Testing:** Preparing for a cyber incident involves more than just a theoretical plan. Like a fighter training under realistic conditions, incident response plans should be stress-tested through simulations or tabletop exercises. These exercises reveal potential weaknesses and prepare teams to handle high-pressure situations, like taking a punch and staying on one's feet.
- **Muscle Memory:** Incident response teams, like fighters, need practice to develop muscle memory. Regular drills ensure that critical steps become second nature. This preparation means that even when things go off-script, responders know how to react and can make quick, effective decisions without losing momentum.
- **Preparation for Unexpected Scenarios:** A good incident response plan considers unpredictable scenarios just as a chess player anticipates their opponent's next surprise moves. Building contingencies for various "what if" situations — such as a simultaneous DDoS attack or ransomware spread — helps a team handle unexpected challenges and minimize panic.
- **Recovery and Resilience:** The most effective response is often not about preventing every hit but recovering quickly. A cyber response plan focuses on fast recovery and resilience, like a boxer training to get back up after a hard blow. This includes precise containment, eradication, and restoration steps, ensuring the organization can recover swiftly and learn from the experience. Cyber incident response planning is about preparing for the unexpected and training to respond effectively, knowing that actual incidents rarely follow a clear plan. Resilience, adaptability, and preparation for the worst-case scenario make a response team "punch-proof" in the cyber arena.

Therefore, it stands to reason that the Chief Information Security Officer (CISO) knows every intricate detail of the organization's IRP, which is beneficial and critical to effective leadership in a crisis. The CISO's intimate knowledge of the IRP enables swift, organized, and effective responses to incidents, minimizing financial, operational, and reputational damage.

This chapter further dissects the critical components of practical plans and best practices for ensuring preparedness for any security incident.

Why Detailed Knowledge of the Incident Response Plan is Essential for CISOs

The CISO's role is increasingly strategic, encompassing risk management, regulatory compliance, and executive communication. An incident response plan is at the heart of an organization's cybersecurity resilience, providing a structured approach to handling cyber incidents. For CISOs, knowing every detail of the IRP enables them to:

- **Lead Confidently in a Crisis:** During an incident, the organization looks to the CISO for clear guidance and assurance. When a CISO knows the IRP inside and out, they can make informed, timely decisions that reassure stakeholders and ensure a coordinated response.
- **Ensure Regulatory Compliance and Legal Preparedness:** Cyber incidents often involve regulatory scrutiny. A deep understanding of the IRP allows CISOs to ensure that responses align with legal and regulatory requirements, reducing liability and protecting the organization's reputation.
- **Protect Organizational Resilience:** The faster and more efficiently an incident is contained and resolved, the quicker the organization can return to normal operations. A CISO who is well-versed in the IRP can anticipate and mitigate disruptions, ensuring the business remains resilient.

Core Components of an Effective Incident Response Plan

A comprehensive IRP covers the entire lifecycle of an incident, from preparation through post-incident review. Mastering these components enables the CISO to lead with precision at every stage. Below, we explore each element in detail.

1. **Preparation** - Preparation is the foundation of effective incident response, focusing on proactive steps to strengthen the organization's readiness for potential threats.

   - **Asset and Data Inventory:** Knowing what needs protection is critical. CISOs should ensure the IRP includes a current inventory of assets, data types, and dependencies, as this helps prioritize protection and inform response actions during an incident.

   - **Roles and Responsibilities:** A clear delineation of roles ensures that each team member knows their responsibilities. The CISO should understand who is involved at every stage of an incident, from technical responders to communications and legal advisors, to ensure the right people are engaged immediately.

   - **Training and Simulation Exercises:** Regular training and tabletop exercises ensure the IR team is prepared. CISOs should understand the frequency and structure of these exercises and track results to identify areas for improvement.

2. **Identification** - The identification phase focuses on detecting, reporting, and verifying incidents. CISOS must understand how the organization detects incidents and determines their validity.

   - **Detection Tools and Technologies:** The CISO should be familiar with all detection tools, such as Security Information and Event Management (SIEM) systems, Intrusion Detection Systems (IDS), and anomaly detection. Understanding the strengths and limitations of each tool enables the CISO to gauge the accuracy of alerts.

- **Incident Classification and Severity:** The IRP should establish criteria for classifying incidents by severity. CISOs should know these criteria to ensure incidents are escalated appropriately, particularly those that require Board or regulatory notification.
- **Early Notification and Escalation Protocols:** Clear notification channels for early alerts are essential. CISOs should know how quickly alerts are expected to reach critical responders and what escalation protocols exist to engage senior management when needed.

3. **Containment** - Containment is the process of stopping an incident from spreading and limiting its impact. Knowing the intricacies of containment strategies is essential for the CISO, as this phase can significantly influence the overall effect of an incident.

- **Short-Term vs. Long-Term Containment Strategies:** Short-term strategies, such as isolating infected systems, help contain immediate damage, while long-term strategies ensure the threat is entirely eradicated. The CISO must understand the details to balance quick action with lasting protection.
- **Decision-Making Protocols for Isolation:** Containment often requires quick decisions about system isolation or network shutdowns. The CISO should know the hierarchy and authority for making these decisions and understand their impact on business operations.
- **Forensic Preservation:** Evidence must be preserved for investigation and possible legal proceedings. CISOs should be familiar with forensic protocols to ensure the

IR team preserves evidence properly, which can help in root cause analysis and legal defense.

4. **Eradication** - Eradication involves removing the threat from the environment and addressing the vulnerabilities that led to the incident. For the CISO, understanding this phase is crucial for immediate remediation and long-term resilience.

   - **Root Cause Analysis:** Eradication relies on understanding the root cause of the incident, whether it's a malware infection, a misconfiguration, or an insider threat. The CISO should ensure that effective root cause analysis methods are in place to guide the eradication process.

   - **Patch Management and Vulnerability Remediation:** If the incident was caused by a specific vulnerability, patching or remediating this weakness is essential. CISOs should understand the organization's vulnerability management process to make sure that similar issues are prevented.

   - **Validation of Eradication Efforts:** After initial eradication, systems should be checked to confirm that the threat is completely removed. CISOs should know what validation measures are in place to verify that affected systems are clean before moving to recovery.

5. **Recovery** - The recovery phase is restoring normal operations and monitoring for any lingering threats. The CISO's detailed knowledge of recovery protocols ensures a seamless return to business as usual.

   - **System Restoration and Testing:** Systems should be restored methodically to avoid reintroducing vulnerabilities. The CISO should understand the sequence and validation procedures for restoring systems, including testing to confirm functionality and security.

- **Stakeholder Communication:** Communication with internal and external stakeholders during recovery is crucial for managing expectations and ensuring transparency. The CISO should know the communication protocols and understand when to engage executives, board members, and potentially affected parties.
- **Post-Recovery Monitoring:** Post-recovery monitoring ensures no residual threats remain in the environment. The CISO should know the duration and scope of this monitoring phase, including specific metrics or alerts to watch for potential reinfections.

6. **Lessons Learned -** The final phase, lessons learned, involves analyzing the incident to enhance future response capabilities. CISOs who understand this phase can leverage incidents as learning opportunities for continuous improvement.

- **Post-Incident Review Meetings:** The CISO should lead post-incident reviews to analyze response effectiveness and identify areas for improvement. Documenting these discussions helps refine the IRP and strengthens the organization's resilience.
- **Updating the IRP:** Insights gained from each incident should lead to updates in the IRP, incorporating new threats, vulnerabilities, and best practices. A CISO who knows this process in detail can ensure the IRP remains relevant and actionable.
- **Cross-Functional Engagement:** By involving stakeholders from across the organization in the lessons-learned phase, the CISO can foster a security-aware culture and promote shared responsibility for resilience.

The Benefits of a CISO's In-Depth Knowledge of the IRP

Knowing the IRP in detail provides significant advantages to both the CISO and the organization, especially during high-stakes incidents.

- **Effective Decision-Making Under Pressure:** With a thorough understanding of the IRP, CISOs can make quick, informed decisions, minimizing confusion and reducing delays during critical phases of an incident.
- **More vital Collaboration Across Departments:** Mastery of the IRP enables the CISO to work seamlessly with other departments, from legal and public relations to HR and compliance, ensuring a well-coordinated response.
- **Increased Trust from Stakeholders:** When CISOs demonstrate expertise in incident response, they build trust among executives, board members, and regulators, establishing Cybersecurity as a critical aspect of organizational strategy.
- **Enhanced Regulatory Compliance:** By mastering the IRP, CISOs can ensure that all incident responses comply with regulatory requirements, minimizing the risk of fines, penalties, and reputational harm.

Best Practices for CISOs in Mastering the Incident Response Plan

- **Regularly Review and Update the IRP:** As cyber threats evolve, so should the IRP. CISOs should conduct regular reviews to ensure the plan reflects current threats, technologies, and regulatory requirements.

- **Conduct Simulations and Tabletop Exercises:** Leading hands-on simulations helps the CISO and the incident response team practice and refine their skills, making responses smoother and more effective during actual incidents.

- **Engage with Cross-Functional Teams:** Incident response is a team effort. CISOs should work with legal, PR, HR, and other teams to ensure that every aspect of the IRP is cohesive and actionable.

- **Document and Integrate Lessons Learned:** CISOs should document lessons learned after each incident or simulation and incorporate these insights into the IRP. Continuous improvement helps strengthen the response framework.

- **Build Relationships with External Partners:** Sometimes, incidents require specialized expertise from forensic analysts, legal advisors, or communications specialists. Establishing relationships with trusted vendors and partners ensures that these resources are readily available when needed.

For a CISO, knowing the intricate details of the incident response plan is not just a matter of preparedness but a foundation for effective leadership in the face of adversity. A CISO who masters the IRP can guide the organization through any security incident with confidence, agility, and clarity, ensuring that cyber threats are contained swiftly and minimizing harm to the organization's assets, operations, and reputation. In Cybersecurity, where every second counts, a CISO's comprehensive knowledge of the IRP is a strategic asset that fortifies resilience, builds trust, and sets the standard for excellence in security leadership.

Chapter 15

KNOW THY FINANCES

Beyond The Code: The Importance Of Financial Acumen For CISOs

As CISOs, we are often so immersed in the technical intricacies of Cybersecurity that we sometimes overlook a crucial skill necessary to be effective at the executive table: financial acumen. Throughout my career—spanning military service, government positions, and leadership roles across multiple sectors, one recurring theme has become abundantly clear: understanding the financial implications of security decisions is critical not only to protect the organization but also to thrive in the role of a modern security leader.

Cybersecurity is a Business Enabler

Many CISOs, myself included in my earlier years, view their role through the lens of technical stewardship. We ensure the fire-walls are configured correctly, incident response plans are air-tight, and threat intelligence feeds are continuously monitored. Over time, however, the realization that Cybersecurity is more than just a technical safeguard; it is a business enabler has become central to how CISOs view ourselves and our roles.

Every decision we make impacts the bottom line. Whether we're deciding on a new security tool, implementing a new data privacy regulation, or calculating the potential impact of a data breach, our actions have financial consequences. The CISO is no longer just a protector of systems; we are integral to maintaining business continuity, safeguarding revenue streams, and driving long-term growth by ensuring trust and resilience. However, we can only achieve this if we can speak the language of finance.

Bridging the Gap Between Security and Business

I've seen the challenge of aligning cybersecurity strategies with business objectives in various industries. In healthcare, for exam-ple, the need to balance the need for stringent security controls with the necessity of operational efficiency and patient care is criti-cal. The financial sector was a delicate dance between regulatory compliance and protecting customer data while ensuring that secu-rity investments made fiscal sense.

Understanding financial principles helped me communicate more effectively with other executives in each instance. Knowing how to present security initiatives regarding Return on Investment (ROI), Total Cost of Ownership (TCO), and risk-adjusted cost-ben-efit analyses enabled me to frame cybersecurity in a way that res-onated with the CFO and the various Boards of Directors I've been privileged enough to report into over the years. When cybersecurity initiatives are presented with a clear financial lens, gaining approval for budgets, resources, and even strategic shifts that align with the company's long-term goals becomes more accessible.

The Stanford Experience: A Game Changer

One of the most transformative experiences in this regard was attending Stanford's "Finance for Non-Financial Managers" course. As someone with a solid technical background but limited formal financial training, I found this course a game-changer. It was eye-opening to see how economic principles could be applied to our world of Cybersecurity. The instructors integrated complex financial concepts into digestible insights that immediately resonated with me. Understanding how to read financial statements, analyze investments, and evaluate business performance through an economic lens empowered me to elevate my contributions as a CISO. For example, investing in a robust security infrastructure may seem expensive upfront. However, considering the cost of a breach, the potential regulatory fines, and the loss of trust from customers, that investment often becomes a sound financial decision. The ability to articulate this in terms that finance and business leaders understand is crucial to driving informed decision-making.

Financial Planning and Analysis (FP&A) is a critical business function that drives the company's budgeting, forecasting, and financial decision-making processes. For CISOs, having a clear understanding of the FP&A process is crucial—not only for managing their budgets but also for aligning cybersecurity goals with broader business objectives and demonstrating the value of security investments to the organization.

The FP&A process provides insights into how the organization allocates resources across departments. By understanding it, CISOs can better navigate the budgeting process, ensuring cybersecurity initiatives receive the necessary funding. Because cybersecurity needs are not static, CISOs must anticipate long-term investment needs, and corporate FP&A processes enable CISOs to plan multi-year strategies, aligning cybersecurity projects with the organization's financial roadmap. Understanding FP&A allows CISOs to quantify and justify the economic impact of security initiatives, helping them demonstrate how cybersecurity investments support revenue growth, risk reduction, and operational efficiency. Critical aspects of FP&A that CISOs should familiarize themselves with include:

- **Budgeting Cycles and Processes:** Each organization typically has nuanced budgeting cycles and associated processes that CISOs must get very intimate with in order to be successful.
- **Annual Budgeting:** Many organizations follow a yearly budgeting process, during which department heads, including the CISO, submit budget proposals. Understanding the budgeting cycle allows CISOs to align cybersecurity requests with this timeline, ensuring funding for critical projects.
- **Quarterly Adjustments:** Some organizations adjust their budgets quarterly to adapt to changing needs or priorities. For CISOs, these adjustments may offer opportunities to request additional funding for new security needs or respond to emerging threats.

Forecasting and Scenario Analysis

- **Predicting Future Security Needs:** Forecasting is a critical component of FP&A, involving revenue, expenses, and resource requirements predictions. CISOs can use forecasting to anticipate future security needs, such as compliance costs, technology upgrades, or increased staff requirements.

- **Scenario Planning:** Scenario analysis helps organizations plan for potential financial outcomes. CISOs can benefit by using scenario planning to predict the economic impact of different cybersecurity scenarios, such as a data breach, regulatory fine, or new compliance mandate.

Cost-Benefit Analysis

- **Evaluating Security Investments:** FP&A teams often conduct cost-benefit analyses to assess potential investments. Understanding this process allows CISOs to frame cybersecurity expenditures in terms of ROI, explaining how investments reduce risk and potential costs associated with breaches or downtime.
- **Prioritizing Projects:** Cost-benefit analysis helps CISOs prioritize initiatives based on business value. Projects demonstrating a solid ROI, such as initiatives to reduce compliance costs or prevent high-risk threats, can gain approval more quickly.

Variance Analysis

- **Tracking Budget vs. Actuals:** Variance analysis compares budgeted expenditures to actual spending. CISOs can use variance analysis to track cybersecurity spending, identify cost-saving opportunities, and adjust budgets based on actual needs.
- **Justifying Unexpected Expenses:** If a significant security incident or unexpected costs arise, variance analysis provides a framework for explaining these expenses to FP&A and finance teams. Benefits of Aligning Cybersecurity Strategy with FP&A

When CIOs integrate their understanding of FP&A into cyber-security planning, it brings significant benefits to both the security team and the organization as a whole:

- **Enhanced Collaboration with Finance and Leadership:** By understanding the FP&A process, CISOs can better communicate with finance teams, executive leadership, and the Board. They can articulate cybersecurity needs in finan-cial terms, improving collaboration and ensuring Cyberse-curity remains a priority. Furthermore, this understanding helps bridge the gap between finance and IT, leading to more substantial cross-functional alignment and more cohe-sive strategies.
- **Efficient Use of Resources:** A solid grasp of FP&A allows CISOs to allocate resources efficiently. By aligning their budget with the organization's financial priorities, CISOs can optimize spending, prioritize high-impact initiatives, and avoid budget cuts that might jeopardize security.
- **Better Risk Management and Business Continuity:** Un-derstanding FP&A enables CISOs to address the financial impact of cybersecurity risks proactively. By quantifying potential costs related to data breaches, regulatory fines, or system downtimes, CISOs can make the case for preventive measures that support business continuity.
- **Strategic Planning and Long-Term Security Roadmap:** The FP&A process involves short- and long-term financial planning. By participating in this process, CISOs can de-velop a multi-year security roadmap aligned with the orga-nization's business goals, allowing for strategic investments in infrastructure, talent, and technology.
- **Improved ROI and Demonstrable Business Value:** Un-derstanding FP&A allows CISOs to measure cybersecurity initiatives' Return on Investment (ROI), helping them jus-tify security spending. By framing cybersecurity in terms of

cost savings, risk reduction, and productivity gains, CISOs can better demonstrate the business value of security.

Practical Steps for CISOs to Engage with FP&A

To effectively leverage FP&A in cybersecurity planning, CISOs can take the following steps:

- **Build a Relationship with the FP&A Team:** Develop a working relationship with the FP&A team to understand financial priorities, budget cycles, and reporting structures. Regular communication with FP&A provides CISOs with insights into the organization's financial strategy and allows for early collaboration on cybersecurity funding.
- **Understand the Organization's Financial Goals and Priorities:** Familiarize yourself with the company's financial objectives. If the company is focused on reducing operational costs or scaling growth, align cybersecurity strategies to support these goals.
- **Develop a Clear Cybersecurity Budget Proposal:** Present a cybersecurity budget that aligns with the company's financial priorities. Outline key initiatives, explain their costs, and emphasize the expected ROI. Use financial terms and metrics, such as cost avoidance, risk reduction, and efficiency gains, to make a compelling case for funding.
- **Leverage Data and Metrics to Justify Security Spending:** Use metrics to quantify the potential financial impact of cybersecurity incidents, such as the cost of a data breach, regulatory fines, or productivity loss. Provide data-driven forecasts to support your budget requests and justify investments.
- **Prepare for Cost-Benefit and ROI Analysis:** Be ready to present a cost-benefit analysis for significant investments,

demonstrating how each initiative aligns with business goals. For example, if investing in a security technology reduces compliance costs, quantify those savings to show the financial benefit.

- **Participate in Scenario Planning:** Work with FP&A to develop cybersecurity-related scenarios, such as the potential financial impact of a breach or a new regulatory requirement. Scenario planning helps the organization understand the economic implications of cyber risks and supports proactive investment in security.

- **Report on Security Metrics and Financial Outcomes:** Use financial metrics to report on the success of cybersecurity initiatives. Presenting security metrics alongside financial outcomes (e.g., reduction in incident costs or regulatory penalties) demonstrates the tangible impact of cybersecurity on the business.

Knowledge of FP&A empowers CISOs to plan budgets strategically, communicate the financial value of cybersecurity, and align security initiatives with broader business goals. By bridging the gap between finance and cybersecurity, CISOs can effectively advocate for the resources needed to protect the organization and demonstrate the return on investment of their efforts. In the modern enterprise, where security is a business enabler, understanding FP&A is more than a financial skill—it's a strategic imperative for CISOs to drive cybersecurity forward.

Chapter 16

KNOW THY BOARD

Navigating Relationships with the Board of Directors, Board Audit & Risk, and Technology Committees

For today's CISO, cybersecurity is no longer a back-office concern but a boardroom imperative. The sophistication of cyber threats and the consequences of breaches require CISOs to engage with their Board of Directors actively, emphasizing the Audit, Risk, and Technology committees. These relationships are essential to aligning cybersecurity strategy with the organization's risk appetite, business priorities, and regulatory responsibilities. In this chapter, I'll explore the importance of understanding these board dynamics, share insights on how CISOs can build effective relationships with key board members, and outline the benefits of becoming a strategic advisor rather than just a technical overseer.

Why Understanding the Board of Directors is Essential for CISOs

CISOs often find themselves walking a fine line between defending against technical threats and navigating corporate politics. Cybersecurity has become a high-priority agenda item in recent years, and the Board expects CISOs to bring clarity to complex cyber issues. However, understanding the Board goes beyond presenting security metrics; it requires grasping the personalities, priorities, and dynamics within the Board.

Each Board is unique regarding risk tolerance, technological knowledge, and corporate strategy. Boards are often composed of experienced professionals, frequently older and less versed in modern technology, with diverse backgrounds in finance, law, operations, and sometimes technology. Recognizing these elements allows CISOs to tailor their presentations, provide context for critical security initiatives, and position themselves as strategic partners. This approach helps secure the budget and buy-in for cybersecurity programs and builds a more proactive and responsive organizational security culture.

The Role of the Audit and Risk Committees in Cybersecurity

Audit Committee - The audit committee's focus on financial integrity, compliance, and internal controls makes them a key ally for CISOs. Given cybersecurity incidents' significant regulatory and economic impact, the audit committee views cyber risks as intertwined with the company's financial stability.

Their concerns typically center around the following:

- **Regulatory Compliance:** They monitor the organization's adherence to regulatory frameworks, such as the SEC's recent cyber disclosure rules, GDPR, or industry-specific mandates like HIPAA and PCI DSS.
- **Incident Response and Reporting:** The audit committee wants assurance that robust processes exist for detecting, responding to, and reporting cyber incidents.
- **Financial Impact of Cybersecurity:** They evaluate the financial implications of cybersecurity budgets and investments, balancing these against overall corporate objectives.

When CISOs work closely with the audit committee, they ensure that cyber initiatives align with the organization's compliance goals and financial controls. A transparent relationship can help the committee understand the business value of security initiatives and foster a risk-aware culture.

Risk Committee - The risk committee focuses on enterprise-wide risk management, balancing strategic objectives with acceptable levels of risk. They play a critical role in helping CISOs frame Cybersecurity as a component of overall business risk.

Fundamental interests of the risk committee include:

- **Risk Appetite and Tolerance:** They set the tone for how much risk the organization is willing to accept and expect security leaders to align their strategy accordingly.
- **Risk Prioritization and Threat Landscape:** They seek insights into emerging threats and want assurance that the company's risk management strategies are forward-thinking.
- **Business Continuity and Resilience:** They evaluate the organization's capacity to withstand and recover from incidents, placing high importance on resilience planning.

By aligning with the risk committee, CISOs gain insight into the Board's view on risk and can better communicate how cybersecurity investments help mitigate those risks. This partnership reinforces the role of cybersecurity as integral to the company's resilience and competitive positioning.

The Role of the Technology Committee in Cybersecurity

The technology committee is often one of the few board groups with members versed in digital strategy and technical matters, making them valuable allies for CISOs. This committee generally oversees technology-driven innovation, digital transformation, and IT infrastructure. Their concerns include:

- **Technology Strategy Alignment:** They ensure cybersecurity strategies support the company's digital initiatives, such as cloud migration, IoT deployments, and AI integration.
- **Innovation and Security Balancing Act:** They advocate for digital innovation while ensuring security does not become a barrier.
- **Emerging Technology Risks:** They seek to understand how adopting new technologies like generative AI, quantum computing, and IoT introduces new risks and how these are mitigated.

The technology committee's grasp of digital initiatives allows CISOs to have more technical discussions about the architecture, control frameworks, and cybersecurity's role in supporting the business's digital agenda. Engaging with this committee early and frequently helps CISOs proactively address potential cyber risks in transformative projects, positioning security as an enabler rather than a hindrance.

Building Effective Relationships with Key Board Members

Developing relationships with individual board members, especially those on the audit, risk, and technology committees, is a strategic investment. Here are some critical steps to foster strong connections:

- **Know Their Backgrounds:** Research each board member's background, career trajectory, and industry expertise. Understanding their experience informs how you present cybersecurity issues. For example, finance-driven board members may appreciate focusing on financial risk and cost-effectiveness, while those with a tech background may prefer discussions around architecture and threat landscapes.
- **Communicate in Business Terms:** Board members are less interested in technical jargon and more focused on how cybersecurity affects business outcomes. Framing your discussions around business impact, such as potential revenue loss, reputational damage, and compliance risk, resonates more effectively.
- **Provide Real-World Scenarios:** Illustrate your points using real-world examples from similar industries. Mentioning notable incidents (like SolarWinds, the Moveit breach, or the impact of SEC rules on cybersecurity disclosures) makes abstract risks tangible and demonstrates your awareness of industry-wide challenges.
- **Offer Strategic Insights, Not Just Operational Data:** Go beyond tactical updates and share insights on how cybersecurity initiatives align with and support the organization's goals. Position yourself as a strategic advisor who helps the Board make informed decisions.

Preparing for Board Presentations

Board presentations are a high-stakes opportunity for CISOs to demonstrate value and secure buy-in. Preparation is vital, and compelling presentations should:

- **Prioritize the Board's Key Concerns:** Focus on issues relevant to each committee. The audit committee emphasizes compliance and incident response. The risk committee highlights risk posture and resilience. The technology committee discusses support for digital transformation and technology risk.
- **Use Clear, Visual Data Representations:** Present data through visuals that board members can grasp quickly, such as dashboards that show risk levels, threat trends, and return on security investment.
- **Anticipate Questions and Concerns:** Prepare for potential questions by considering the organization's risk profile, recent cyber incidents, and strategic initiatives. Proactively addressing common questions builds credibility.
- **Present Actionable Insights:** Offer actionable recommendations, such as risk reduction strategies or areas where additional investment is needed. Clearly outline how these actions align with the Board's risk tolerance and business objectives.

The Benefits of Effective Board Relationships

By building effective relationships with the Board, particularly the audit, risk, and technology committees, CISOs can realize significant benefits:

- **Improved Resource Allocation:** When boards understand the critical nature of cybersecurity, they're more likely to support necessary investments in personnel, tools, and technology.

- **Enhanced Organizational Culture:** Board-level support for cybersecurity fosters a culture of security throughout the organization, influencing all departments to prioritize cyber risk.
- **Increased Strategic Influence:** Strong relationships with board members enable CISOs to transition from being seen as tactical problem solvers to strategic business partners who support growth and resilience.
- **Proactive Risk Management:** An aligned board-CISO relationship ensures the organization remains proactive in managing risks, addressing emerging threats, and meeting regulatory requirements.

Conclusion

In today's landscape, CISOs must transcend technical expertise and engage the Board of directors as strategic partners. Understanding the nuances of each committee and tailoring your approach to their unique priorities will solidify your role as an indispensable advisor on cyber risk. By fostering these relationships, you elevate your influence and ensure that cybersecurity becomes a cornerstone of the company's success.

Chapter 17

KNOW THY LEADERSHIP STYLE

The Strategic Value of Leadership Styles for CISOs

Know Thy Leadership Style - The Strategic Value of Leadership Styles for CISOs

As CISOs move into strategic roles beyond technical oversight, our effectiveness increasingly depends on leadership abilities. The ability to influence, inspire, and align diverse teams around complex security initiatives requires more than technical expertise; it requires a deep understanding of leadership style. Knowing our leadership style empowers CISOs to harness their strengths, build high-performing teams, and foster a security culture supporting business resilience and growth. In this chapter, we'll explore the importance of leadership styles for CISOs, outline the major styles, and discuss how understanding one's approach can directly impact cybersecurity outcomes.

The Evolving Role of the CISO: From Technical Leader to Strategic Partner

Historically, the role of the CISO was rooted in technology, focusing on managing firewalls, detecting threats, and ensuring regulatory compliance. However, the CISO's role has evolved into a core part of business strategy, requiring a balance between security initiatives and business goals. This shift demands new skills, particularly in leadership, as CISOs now interact regularly with executives, board members, and cross-functional teams.

- **Influence at the Executive Level**: To gain support for cybersecurity investments and initiatives, CISOs must communicate complex risks that resonate with business leaders. Leadership style is critical here, impacting how effectively the CISO can advocate for resources and align security with corporate priorities.
- **Fostering a Security-Conscious Culture**: Building a security-minded organization requires more than policies and protocols; it demands the ability to inspire employees to adopt security as part of their daily routines. CISOs who understand their leadership style can adapt it to foster a culture where cybersecurity is everyone's responsibility.
- **Navigating Change**: The rapidly evolving cyber threat landscape means that CISOs must be able to lead teams through continuous change, often in high-pressure situations. A clear understanding of leadership style can help CISOs motivate their teams, manage stress, and maintain morale during challenging times.

Understanding Leadership Styles and Their Impact on Security Outcomes

Leadership style encompasses the unique ways an individual motivates, guides, and interacts with their team. CISOs can tailor their approach to different scenarios and organizational relationships by understanding their style. Below, we explore some common leadership styles and their implications for cybersecurity.

Transformational Leadership

Transformational leaders inspire and motivate their teams by creating a compelling future vision. They are known for their ability to lead change, encourage innovation, and foster a culture of continuous improvement.

- **Benefits for CISOs**: Transformational leadership is highly effective in cybersecurity, where threats and technologies constantly evolve. This style allows CISOs to encourage innovative problem-solving, which is crucial for staying ahead of attackers. Transformational CISOs inspire loyalty and commitment, creating a robust and adaptable security team.
- **Challenges**: Transformational leadership can be demanding and may risk burnout if not balanced with clear boundaries. CISOs using this style should ensure that the team's drive for innovation is balanced with realistic expectations and resources.

Transactional Leadership

Transactional leaders focus on structure, goals, and rewards. They set clear objectives and provide recognition or corrective feedback based on performance. This style is often effective for maintaining stability and meeting specific targets.

- **Benefits for CISOs**: Transactional leadership can be highly effective in environments where compliance and regulatory standards are critical. It helps ensure that security policies are followed consistently, and that staff are held accountable for meeting established standards. This style can also help manage day-to-day operations and respond to immediate threats.
- **Challenges**: While transactional leadership can be effective for short-term goals, it may stifle creativity and limit team members' willingness to take initiative. CISOs with a transactional approach should consider blending it with transformational elements to foster compliance and innovation.

Servant Leadership

Servant leaders prioritize the needs and growth of their team, empowering them to succeed and supporting their development. This style fosters trust, collaboration, and a sense of community within the team.

- **Benefits for CISOs**: A servant leadership approach is precious in cybersecurity, where trust and collaboration are essential. CISOs who embrace servant leadership tend to build loyal, motivated teams who feel supported and are more likely to stay engaged over the long term. This approach can lead to higher job satisfaction and lower turnover, which is especially valuable in an industry with high demand for talent.

- **Challenges**: While servant leadership promotes team unity, it can sometimes slow decision-making if the CISO places too much emphasis on consensus. CISOs may need to balance this approach with more directive leadership to make quick, decisive actions in high-stakes situations.

Democratic Leadership

Democratic leaders seek input from their team and make decisions based on collective feedback. This inclusive style values collaboration, encouraging team members to contribute ideas and feel invested in outcomes.

- **Benefits for CISOs**: The democratic style fosters open communication, essential for identifying security issues early. By encouraging input, CISOs can leverage their team's diverse perspectives and expertise, leading to well-rounded strategies and a stronger sense of ownership among team members.
- **Challenges**: Democratic leadership can be time-consuming, and in situations that require swift action, it may delay critical decisions. CISOs should use this style judiciously, balancing inclusivity with the need for prompt responses to urgent security threats.

Authoritative (Visionary) Leadership

Authoritative leaders set a clear direction and inspire their team to follow a shared vision. They provide strong guidance and act as role models, making them particularly effective in situations that require decisive leadership.

- **Benefits for CISOs**: In cybersecurity, where a clear vision and decisive action are often needed, an authoritative style can be highly effective. This style enables CISOs to communicate the importance of security initiatives and inspire confidence in their teams. Implementing significant changes, such as a security transformation or crisis response, is precious.

- **Challenges**: While authoritative leadership can be inspiring, it may come across as rigid if team members feel they lack input. CISOs should ensure that they also encourage feedback to keep the team engaged and open to new ideas.

Aligning Leadership Style with Cybersecurity Goals

Understanding one's leadership style allows CISOs to adapt their approach to different situations, enhancing their organizational impact. By aligning their style with specific goals, CISOs can strengthen their influence and improve outcomes in critical areas of cybersecurity.

Building a Resilient Security Team

CISOs are responsible for building and maintaining a resilient security team capable of responding to diverse threats. Knowing their leadership style helps them create an environment where team members feel valued, supported, and motivated to succeed.

- **Transformational and Servant Leadership**: These styles create a positive culture where team members feel empowered and motivated. A CISO who combines these approaches can build a resilient, high-performing team invested in the organization's security mission.
- **Transactional and Democratic Leadership**: These styles work well in managing daily operations and ensuring compliance, particularly in environments where structure and policy adherence are critical. They can be instrumental in

managing regulatory requirements and maintaining consistent security practices.

Promoting Cross-Departmental Collaboration

Cybersecurity is no longer isolated in the IT department; it requires cooperation across departments, from finance to human resources to operations. Leadership style is critical in fostering these relationships and aligning cybersecurity initiatives with business goals.

- **Democratic and Servant Leadership**: CISOs with these styles often excel at building trust and collaboration with other departments, emphasizing open communication and mutual support. This approach helps integrate cybersecurity into the broader organization and fosters a culture of shared responsibility.
- **Authoritative Leadership**: When a clear vision for security is needed across departments, an authoritative CISO can effectively communicate the importance of security initiatives and inspire alignment. This approach is precious for gaining buy-in from senior leaders.

Driving Innovation in Cybersecurity

Innovation is critical in cybersecurity to stay ahead of evolving threats and improve efficiencies. By leveraging their leadership style, CISOs can foster a culture encouraging experimentation and adaptation.

- **Transformational Leadership**: This style is particularly effective for driving innovation, as it encourages the team to think creatively and explore new solutions. Transformational CISOs will likely create a culture where continuous improvement is valued, and innovation thrives.

- **Democratic Leadership**: CISOs who adopt a democratic style can involve team members in brainstorming sessions and strategic planning, which often leads to innovative ideas. CISOs can tap into the team's collective intelligence by empowering employees to contribute.

Adapting Leadership Style to Meet Evolving Challenges

The dynamic cybersecurity landscape requires CISOs to be flexible in their leadership approach. Understanding and adapting one's leadership style allows CISOs to respond effectively to different challenges, from crisis management to long-term strategic planning.

- **Crisis Situations**: During a security incident, an authoritative or transactional approach may be necessary to ensure a swift, organized response. In such situations, CISOs should be prepared to take decisive action, providing clear direction to their teams.
- **Long-Term Initiatives**: Transformational and servant leadership styles are often most effective for initiatives requiring long-term buy-in, such as implementing a new cybersecurity framework or building a security culture. These styles allow CISOs to inspire and engage their teams over time.
- **Adapting to Team Needs**: Different team members may respond better to different leadership styles. By understanding their primary leadership style and remaining flexible, CISOs can adapt their approach to support each team member's development and engagement.

Conclusion: A CISO's Commitment to Workforce Development

Cyber workforce development is more than just a human resources initiative—it's a strategic priority that ensures your organization is prepared to face the threats of today and tomorrow. As CISOs, we must lead by example, fostering a culture of continuous learning, mentorship, and growth within our teams. Through skills assessments, career pathing, and workforce development, we ensure that our teams are equipped not just to react to threats but to anticipate and mitigate them.

By partnering with organizations like Year Up, NPower, Per Scholas, the military, and local colleges, we expand our talent pool and bring new voices and perspectives into cybersecurity. Through conferences and resources like webinars and podcasts, we stay ahead of the curve, ensuring that our strategies are informed by the latest intelligence and trends. Ultimately, the strength of our cybersecurity workforce is the foundation upon which our security programs are built—and as CISOs, we are the architects of that foundation.

Chapter 18

KNOW THY TALENT

The Importance of Cyber Workforce Development for CISOs

T
hroughout my career as a CISO, working across diverse industries—from the military and government to finance and management consulting—I've consistently faced the challenge of building and maintaining a cybersecurity workforce capable of responding to ever-evolving threats.

My approach's cornerstone has been my focus on identifying young talent and nurturing their potential through a structured approach of challenging, career-broadening assignments. From the outset, I prioritize finding individuals with strong analytical abilities, a dedication to learning, and a natural aptitude for leadership. Once identified, I invest in their growth by assigning them to complex projects that push their technical and strategic thinking. I emphasize real-world exposure to critical aspects of cybersecurity—from threat intelligence to incident response—so they gain a well-rounded skill set. Recognizing the importance of developing technical skills and critical executive skills, I involve them in board presentations, cross-functional team meetings, and high-stakes discussions with business leaders, allowing them to refine their communication and decision-making abilities. My approach includes continuous mentorship, feedback, and, importantly, encouraging them to take ownership of their projects, fostering accountability and confidence.

Our talent pool is stretched thin, and the demand for skilled cybersecurity professionals continues to grow. As CISOs, we're not only responsible for defending our organizations from cyber threats but also for developing and nurturing the teams that do the defending. Cyber workforce development is no longer a nice-to-have; it's a strategic imperative that touches every aspect of security leadership—from skills assessment and career pathing to training, mentoring, and the creation of talent pipelines.

Skills Assessment: The Starting Point

Before effectively developing a cyber workforce, you need to know where you stand. Skills assessment is critical in identifying the strengths and weaknesses within your team. Over the years, I've found that too many organizations rely on outdated job descriptions and assumptions about their teams' knowledge. In an environment where cyber threats change by the day, your team's skills need to evolve in lockstep.

In healthcare, for example, the focus might be on safeguarding patient data under HIPAA, while in finance, the emphasis is on protecting financial transactions and maintaining regulatory compliance. The first step toward building a resilient security team in every industry I've worked in has always been conducting comprehensive skills assessments. These assessments help you identify the technical and soft skills needed—such as critical thinking, communication, and problem-solving—that are just as important in cybersecurity.

Career Pathing: Guiding Growth

Career pathing becomes the next critical step once you understand your team's capabilities. One of the mistakes I've seen too often in cybersecurity is the lack of a clear path to progression. Without it, even the most talented individuals can stagnate, leading to high turnover and frustration. Career pathing provides a roadmap for individuals to grow, learn, and advance within the organization.

I've found that when you give your people a clear vision of their future—whether progressing from a junior analyst role to a senior engineer or moving into leadership—retention improves, and engagement rises. Importantly, career pathing isn't just about promotions. It's also about lateral moves, allowing individuals to expand their skills across different domains—cloud security, threat intelligence, or incident response. This flexibility helps build a well-rounded team and prepares your organization for future challenges.

Workforce Development: Continuous Learning

Workforce development needs to be continuous, dynamic, and proactive. Threats evolve, regulations change, and new technologies emerge, so investing in regular, ongoing training is essential. In the energy sector, for example, the focus might be on securing critical infrastructure, while in healthcare, it could be on protecting electronic health records. Training programs should be tailored to your industry's unique needs and forward-looking enough to prepare your team for future developments.

For years, I've implemented development programs that focus on real-world applications. Simulated attacks, red team/blue team exercises, and hands-on labs have been invaluable tools in preparing my teams for the complexities of the modern threat landscape.

At the same time, I encourage participation in conferences like RSA, Black Hat, and InfoSec World. These events offer cutting-edge insights into the latest threats, technologies, and networking opportunities that can lead to future collaborations. Additionally, industry podcasts are a powerful tool for staying current—conversations with experts, new ideas, and emerging trends are delivered directly to the team and me regularly, helping us stay ahead of the curve.

Mentoring: Building Tomorrow's Leaders

Mentorship is an integral part of cyber workforce development, which I've always prioritized throughout my career. For a team to thrive, they need more than technical training; they need guidance, support, and the opportunity to learn from those who have already navigated the path ahead. Mentoring plays a crucial role in not just transferring knowledge but also shaping the culture of your team.

I've been fortunate enough to mentor many individuals who have become leaders in their own right. These relationships are not one-sided; mentoring also provides value to the mentor, helping me stay grounded, maintain perspective, and foster a collaborative culture within the team. A successful mentoring program can transform a good team into a great one, ensuring that knowledge and experience are continuously passed down to the next generation of cybersecurity professionals.

Talent Pipeline Development: Tapping Into New Resources

To address the talent shortage in Cybersecurity, CISOs must proactively create and nurture a talent pipeline. We can no longer afford to rely solely on traditional hiring practices, especially given the rapid pace at which cyber threats evolve. Throughout my career, I've partnered with organizations like Year Up, NPower, and Per Scholas—all of which focus on providing training to underrepresented communities, veterans, and individuals seeking to pivot into cybersecurity from other careers. These partnerships are essential in diversifying our talent pool and bringing fresh perspectives to our teams.

Additionally, the military has always been an excellent source of talent for cybersecurity. Veterans often come equipped with a unique blend of discipline, problem-solving, and technical skills that translate well into cybersecurity. I've seen firsthand how military experience can accelerate the development of cybersecurity professionals, and I've had great success in hiring veterans into critical roles within my teams.

Another critical component of pipeline development is partnering with local colleges and universities. Creating internship and apprenticeship programs has been one of the most effective strategies I've used to bring in new talent. These programs provide students with hands-on experience, expose them to real-world challenges, and allow us to assess their fit for long-term roles. Many apprentices and interns I've mentored have become full-time team members, contributing valuable insights and energy to our cybersecurity operations.

Staying ahead in cybersecurity requires continuous learning, and one of the best ways to do this is by engaging with the broader cybersecurity community through webinars, conferences, and podcasts. I've always attended conferences because they provide a platform for CISOs and other security professionals to exchange ideas, learn about the latest threats and innovations, and hear from experts across the industry. These conferences often introduce new tools and strategies I can bring back to my team, ensuring we remain agile and responsive to the changing landscape.

Webinars and podcasts have become an essential resource for staying current. In the fast-moving world of cybersecurity, listening to experts break down the latest incidents, vulnerabilities, and innovations in real time has been invaluable. I encourage my teams to engage with these resources regularly, as they offer insights into the challenges and successes other organizations face, helping us adapt and improve our practices.

Conclusion: A CISO's Commitment to Workforce Development

Cyber workforce development is more than just a human resources initiative—it's a strategic priority that ensures your organization is prepared to face the threats of today and tomorrow. As CISOs, we must lead by example, fostering a culture of continuous learning, mentorship, and growth within our teams. Through skills assessments, career pathing, and workforce development, we ensure that our teams are equipped to react to threats and anticipate and mitigate them.

By partnering with organizations like Year Up, NPower, Per Scholas, the military, and local colleges, we expand our talent pool and bring new voices and perspectives into cybersecurity. Through conferences and resources like webinars and podcasts, we stay ahead of the curve, ensuring that the latest intelligence and trends inform our strategies. Ultimately, the strength of our cybersecurity workforce is the foundation upon which our security programs are built—and as CISOs, we are the architects of that foundation.

THE FUTURE OF CISOS: OPPORTUNITIES IN LEADERSHIP, COMPLIANCE, AND CYBER RESILIENCE

Opportunities in Leadership, Compliance, and Cyber Resilience

The future for Chief Information Security Officers (CISOs) is full of promise as organizations increasingly recognize the essential role of cybersecurity in corporate strategy. Today's CISO is no longer confined to managing technical defenses but is crucial to business resilience, regulatory compliance, and corporate governance. Recent developments, such as the U.S. Securities and Exchange Commission's (SEC) new cyber ruling, are shaping a future where cybersecurity sits at the center of enterprise priorities. This chapter will explore the positive opportunities these trends present for CISOs, focusing on the SEC's ruling, the emerging role of CISOs in boardrooms, and the new pathways to leadership and influence within organizations.

The SEC's Cybersecurity Disclosure Ruling: Elevating the CISO's Role

In July 2023, the SEC introduced groundbreaking cybersecurity regulations requiring public companies to disclose cyber incidents and their cybersecurity practices more transparently. This ruling represents a positive development, as it reinforces the importance of cybersecurity in corporate governance and places CISOs in a central role in regulatory compliance and risk management. The SEC ruling highlights how essential cybersecurity is to business health and opens doors for CISOs to demonstrate the value of their work to executives and the Board.

Essential Requirements of the SEC's Cyber Ruling

- **Timely Disclosure of Material Cyber Incidents**: Public companies must report material cyber incidents within four business days, provided they have a significant impact on financial, operational, or reputational health. This means CISOs will be pivotal in assessing incidents and guiding disclosure decisions.
- **Annual Disclosure on Cybersecurity Practices and Board Oversight**: Companies must now provide annual details on their cybersecurity risk management and governance. This includes disclosing the Board's role in overseeing cybersecurity and, where relevant, detailing the cybersecurity expertise present within the Board.

Opportunities Created by the SEC Ruling

The SEC's cyber ruling brings new opportunities for CISOs to showcase their contributions to organizational resilience, build strong relationships with the Board, and enhance their leadership role within their companies.

- **Enhanced Visibility and Influence**: The SEC's new disclosure requirements position CISOs as crucial in boardroom discussions, providing regular updates on cybersecurity risks and ensuring board members are informed and engaged in risk management. This visibility allows CISOs to highlight their teams' successes, demonstrate ROI for cybersecurity investments, and align security initiatives with business goals.
- **Clear Pathways for Career Growth**: As the SEC ruling calls for increased cybersecurity expertise at the board level, experienced CISOs have opportunities to serve as advisors or even join boards themselves. Their specialized knowledge becomes invaluable as boards strive to comply with cybersecurity governance expectations, opening doors for CISOs to expand their career horizons.
- **Empowering CISOs as Strategic Advisors**: The SEC ruling encourages CISOs to go beyond technical security discussions and align cybersecurity with broader business objectives. This shift allows CISOs to build on their strategic impact, working alongside executives to integrate cybersecurity into digital transformation, risk management, and growth initiatives.

Best Practices for Leveraging the SEC Ruling

- **Strengthen Incident Response Frameworks**: Prepare for timely and accurate disclosures by establishing robust incident response protocols with clear criteria for assessing materiality and reporting processes.
- **Foster Board Engagement**: Regularly communicate with board members to provide clear, strategic insights on cyber risks, opportunities, and the evolving threat landscape. Use this interaction to build solid and trusted relationships with the Board.
- **Promote Transparency and Accountability**: Demonstrate the CISO team's commitment to transparency by documenting cybersecurity strategies, risk assessments, and incident responses. This approach builds confidence among stakeholders and establishes the CISO as a proactive leader.

CISO Liability: A Path to Greater Responsibility and Recognition

As CISOs gain prominence in corporate governance, some have raised concerns about personal liability, especially following high-profile cases where security executives were held accountable for their organization's cybersecurity practices. However, this trend offers positive implications, as it reflects the growing importance of cybersecurity within the business and the acknowledgment of the CISO as a senior executive on par with the CFO, COO, and other C-suite roles.

Types of Potential Liabilities and How They're Changing the CISO Role

- **Accountability for Cyber Risk Management**: With the SEC ruling, CISOs are responsible for ensuring accurate and thorough disclosures about cyber risk. This responsibility positions the CISO as an executive directly influencing the company's risk profile, reinforcing its role in safeguarding business interests.

- **Board-Level Responsibilities and Corporate Governance**: Increasingly, boards invite CISOs to discuss cybersecurity policies, risks, and strategies. This involvement allows CISOs to guide board-level decisions on technology and cyber risks, elevating their role as primary advisors in areas critical to business success.

Opportunities for Growing CISO Accountability

The increased focus on CISO accountability can be leveraged positively, positioning the CISO as a central figure in corporate governance and risk mitigation.

- **Executive Empowerment**: Greater accountability comes with the authority to make decisions that impact company-wide cybersecurity. This shift allows CISOs to advocate more effectively for necessary resources, policies, and investments, driving improvements that benefit the organization's resilience.

- **Enhanced Recognition and Compensation**: As CISOs take on responsibilities that carry potential liabilities, companies recognize this level of commitment with competitive compensation, expanded benefits, and, increasingly, liability protections. This recognition reflects the value of CISOs as critical leaders and encourages a shift toward robust, resilient security practices.

- **Growth in Professional Development Opportunities**: To manage potential liabilities effectively, CISOs are encouraged to pursue advanced training and certifications in governance, risk management, and regulatory compliance. These efforts support their professional development and make them more valuable leaders within their organizations.

Best Practices for Thriving with Increased CISO Accountability

- **Embrace Transparent Communication**: Ensure transparent and open communication with executives, stakeholders, and board members. A transparent and proactive approach builds trust and fosters collaboration around cybersecurity.
- **Invest in Legal and Regulatory Knowledge**: Stay informed about evolving cyber regulations, liability protections, and industry standards. Gaining expertise in these areas will empower CISOs to navigate new responsibilities confidently.
- **Promote Cybersecurity as a Shared Responsibility**: Advocate for a culture where cybersecurity is seen as everyone's responsibility, from the boardroom to the frontline. By engaging other leaders and employees, CISOs can foster a collaborative approach to security that mitigates organizational risks.

Cyber Insurance as a Strategic Asset for CISOs

As cyber threats become more costly and complex, cyber insurance has emerged as a strategic tool to manage risk, providing financial protection against data breaches, ransomware attacks, and regulatory penalties. However, the cyber insurance landscape is evolving, with insurers requiring more rigorous security controls. For CISOs, this trend presents an opportunity to build a stronger case for cybersecurity investments and further reinforce the organization's resilience.

Positive Trends in Cyber Insurance for CISOs

- **Elevated Security Standards Drive Improved Security Posture**:
- Insurers increasingly require companies to implement strong security measures, such as Multi-Factor Authentication, Endpoint Detection, and Response (EDR) systems. CISOs can leverage these requirements to secure budget approval for security initiatives that enhance the overall security posture.
- **Integration of Insurance into Incident Response**: Many cyber insurance policies now include provisions for post-incident support, such as forensic analysis, crisis management, and legal counsel. This support can streamline the response process, allowing CISOs to focus on critical response tasks with expert assistance.
- **Enhanced Business Continuity**: Cyber insurance provides a financial safety net and helps sustain business operations in the face of disruptions. By including business interruption coverage, cyber insurance helps CISOs minimize the impact of incidents on revenue, reputation, and stakeholder confidence.

Best Practices for Leveraging Cyber Insurance Effectively

- **Align Cyber Insurance with Risk Management Goals**: Evaluate insurance coverage based on the company's specific risk profile, ensuring policies adequately protect high-risk areas, such as ransomware and data breaches.
- **Demonstrate Compliance with Insurance Standards**: Conduct regular security assessments to demonstrate adherence to insurer standards. By meeting these requirements, CISOs can qualify for favorable rates and potentially reduce premiums over time.
- **Build Relationships with Insurance Providers**: Establish regular communications with insurance providers to align expectations, clarify coverage terms, and ensure the insurer is informed of security improvements and changes in the risk environment.

The Future: CISOs as Strategic Partners in Corporate Resilience

The SEC's cyber ruling, the shift toward CISO accountability, and the evolution of cyber insurance reflect a future where CISOs are recognized as essential strategic partners. These trends empower CISOs to take on expanded roles, contribute to corporate resilience, and secure executive-level influence.

Key Opportunities in the Future of the CISO Role

- **A Place in the C-Suite**: As Cybersecurity becomes integral to corporate governance, more organizations are elevating CISOs to the executive team. This development provides CISOs with direct access to business strategy discussions, enabling them to align cybersecurity more closely with organizational objectives.

- **Expansion into Board Advisory Roles**: With the growing demand for cybersecurity expertise, many CISOs are moving into advisory or board roles. Their insights are invaluable as boards increasingly prioritize cybersecurity, paving the way for CISOs to shape board strategies and gain invaluable experience in governance.
- **Pathways to Greater Leadership Impact**: As CISOs assume greater responsibility, they are positioned to champion digital transformation initiatives and drive innovation. In doing so, they can help organizations balance security with business agility and resilience, turning cybersecurity into a competitive advantage.

CONCLUSION

Amid chaos, there is also
opportunity.
— Sun Tzu.

A s we conclude this journey through the strategies, lessons, and real-world experiences shared in this book, one time-less principle remains at the forefront: **the relevance of Sun Tzu's "Art of War" for modern cybersecurity leadership**. Despite being written over 2,500 years ago, Sun Tzu's strategic insights are as vital today in the digital battlefield as they were on ancient battlefields. For Chief Information Security Officers (CISOs), understanding how to know oneself, the enemy, and the landscape is foundational to building robust and adaptive defenses.

The modern CISO must defend against attacks and predict and anticipate them. This proactive mindset is the key to achieving sustained security, and it's precisely this strategic thinking that makes Sun Tzu's wisdom more relevant than ever.

How To Keep Using This Book

As the cyber landscape shifts, this book should serve as an evolving guide rather than a one-time read. Here's how you can continually use it to stay ahead:

- **Regular Reflection**: Periodically revisit the early chapters on self-awareness, particularly those focused on internal inventory (e.g., knowing your assets, software, and attack surface). As your organization grows or changes, these principles must be re-evaluated. Use the book as a reference to ensure you stay aligned with these core competencies.

- **Strategic Adaptation**: Use the chapters on industry insight and geopolitical awareness as a living document. The threats you face will change as you move through your career or as your organization shifts to industries or geographies. Revisit the lessons on industry-specific challenges and geopolitical threats to adapt your strategies accordingly.

- **Cross-Functional Communication**: The lessons on aligning cybersecurity with business objectives, vendor management, and regulatory compliance will always be crucial. Use these sections as a playbook for how to engage and communicate with other C-suite leaders and boards. The book provides a solid foundation for translating technical challenges into business opportunities.

- **Training and Mentorship**: For seasoned CISOs, this book can also serve as a training manual for developing your next generation of cyber leaders. Chapters on talent management and nurturing the next wave of cyber guardians should be a recurring reference in your leadership toolkit.

- **Evolving with Emerging Technologies**: The chapters that deal with emerging technologies, such as Zero Trust Architecture and securing IoT and cloud environments, are areas that will continuously evolve. As new technologies arise, return to these sections to adjust your strategies, ensuring they remain relevant in the face of rapidly advancing threats and tools.

- **Crisis Preparation**: The various crisis case studies, like the Bangladesh Bank heist and the IRS NIMDA worm incident, should be used as both a training tool and a strategic guide during incident response drills. Practice these scenarios with your team to prepare for similar high-stress situations, knowing that the core principles of response and recovery will remain constant.

Closing Thoughts: A CISO's Lifelong Strategy

In the end, the path of a CISO is not linear. It is a journey marked by constant learning, adaptation, and strategic foresight. The principles laid out in this book—combined with Sun Tzu's teachings—offer a timeless blueprint for navigating this journey. The threats will evolve, the stakes will grow, and the terrain will shift beneath your feet. Yet, the core strategies of self-awareness, preparation, and adaptability will always serve as your compass.

So, as you close this book today, know that this is not an ending but the beginning of a continued journey in cybersecurity leadership. Keep returning to these lessons, applying them, and refining them as the digital battlefield evolves. Let the enduring wisdom of Sun Tzu be your guiding force, reminding you that the best defense is built on deep knowledge, strategic foresight, and relentless preparation.

About the Author

DEVON BRYAN is an award-winning Global Chief Security Officer (CSO) with almost 30 years of experience leading global teams and supporting complex business in various industries, mitigating technology risks while enabling top-line business revenue growth. Devon's career reflects a demonstrated track record of successfully delivering large-scale ($400M+) global cybersecurity management programs that balance cost-effective pragmatic risk-based strategies with complex businesses priorities.

A military veteran of the US Air Force, Devon currently serves as the Global Chief Security Officer (CSO) for the Fortune 200, Booking Holdings Inc where he is charged with developing and implementing the global security strategy that's aligned with Booking Holdings key business priorities. As a strategic enabler of Booking Holdings global technology-dependent business operations, Devon collaborates with senior business executives across the company in the support of growing top-line revenue, driving bottom-line efficiencies all while stimulating innovation, assuring compliance, the protection of personal data and corporate assets, increasing organizational capability and advancing productivity within and across Booking Holdings.

Prior to joining Booking Holdings in January 2025 as the Global Chief Security Officer (CSO), Devon was EVP and Global

Chief Information Security Officer (CISO) for the S&P 500 Carnival Corporation &plc, where he was responsible for establishing and maintaining a comprehensive information security strategy and program to ensure that information assets and technologies of the world's leading cruise line company were appropriately protected. A five-time CISO leading large complex board reportable technology risk management programs, Devon also served as the Chief Information Security Officer (CISO) for MUFG Union Bank (Americas), KPMG North Americas, The US Federal Reserve System (Fed), global outsourced payroll provider ADP Inc and was the Deputy CISO for the Internal Revenue Service (IRS).

www.ingramcontent.com/pod-product-compliance
Lightning Source LLC
Chambersburg PA
CBHW061253220326
41599CB00028B/5634